Genome Informatics 2009

Genome Informatics Series Vol. 22

ISSN: 0919-9454

Genome Informatics 2009

Proceedings of the 9th Annual International Workshop on Bioinformatics and Systems Biology (IBSB 2009)

Boston University, Boston, USA

27 – 29 July 2009

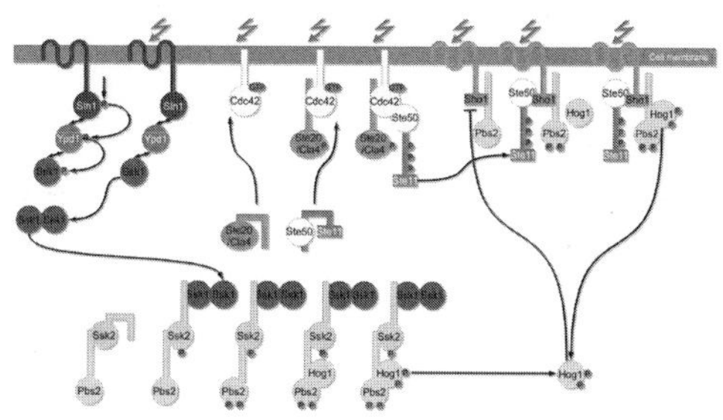

Editors

Charles DeLisi
Boston University, USA

Minoru Kanehisa
Kyoto University, Japan

Edda Klipp
Humboldt-Universität zu Berlin, Germany

Satoru Miyano
University of Tokyo, Japan

Scott Mohr
Boston University, USA

Iwona Wallach
Charité – Universitätsmedizin Berlin, Germany

Imperial College Press

Published by

Imperial College Press
57 Shelton Street
Covent Garden
London WC2H 9HE

Distributed by

World Scientific Publishing Co. Pte. Ltd.
5 Toh Tuck Link, Singapore 596224
USA office: 27 Warren Street, Suite 401-402, Hackensack, NJ 07601
UK office: 57 Shelton Street, Covent Garden, London WC2H 9HE

British Library Cataloguing-in-Publication Data
A catalogue record for this book is available from the British Library.

GENOME INFORMATICS 2009
Genome Informatics Series Vol. 22
Proceedings of the 9th Annual International Workshop on Bioinformatics and Systems Biology
(IBSB 2009)

ISBN-13 978-1-84816-569-4
ISBN-10 1-84816-569-2

Printed by FuIsland Offset Printing (S) Pte Ltd, Singapore

CONTENTS

PREFACE

Genome Informatics Vol. 22 contains a selection of peer-reviewed papers presented at the Ninth Annual International Workshop on Bioinformatics and Systems Biology on 27–29 July of 2009. This time the workshop was held at Life Science and Engineering Building (LSEB) of Boston University, Boston, MA, USA. This international workshop on Bioinformatics and Systems Biology is part of a collaborative educational program involving several leading research institutes: Boston University in the United States, Charité - Universitätsmedizin Berlin, Humboldt-Universität zu Berlin, Kyoto University in Japan and the University of Tokyo. This student-focused event has been held since 2001 to provide doctoral students and young researchers with opportunities to present and discuss their research objectives, approaches and results in the emerging field of genomics, systems biology and bioinformatics. The first workshop was held 2001 in Berlin. It was organized by Prof. Dr. Reinhart Heinrich (1946–2006), a co-founder of this series of workshops. Since 2001, the workshop has been held in Boston (2002), Berlin (2003), Kyoto (2004), Berlin (2005), Boston (2006), Tokyo (2007), and Berlin (2008). The present workshop was organized by the following institutions and programs:

- Boston University Graduate Program in Bioinformatics
- Charité - Universitätsmedizin Berlin
- Freie Universität Berlin
- University of Tokyo Global COE Program – Center of Education and Research for Advanced Genome-Based Medicine
- The International Research Training Group (IRTG) Genomics and Systems Biology of Molecular Networks
- International Research and Training Program on Bioinformatics and Systems Biology, Kyoto University Bioinformatics Center
- Max-Delbrück Center for Molecular Medicine in Berlin
- Max Planck Institute for Molecular Genetics in Berlin
- Max Planck Institute of Molecular Plant Physiology in Potsdam

We first performed the workshop and to collect and review the manuscripts after the workshop, such that the discussions and criticisms at the workshop could be considered appropriately by the authors. However, there was also a pre-selection of the oral and poster contributions to be accepted at the workshop. The contributors were then allowed to submit manuscripts for Genome Informatics. These contributions were reviewed by the members of the workshop event. We have selected 17 papers after revision. These papers will be indexed in Medline, and their electronic versions are freely available from the website of Japanese Society for Bioinformatics as Genome Informatics Online (http://www.jsbi.org/modules/ journal/index.php/index.html). Former publications are also electronically available as Genome Informatics Vol. 15, No. 1 (2004), Vol. 16, No. 1 (2005), Vol. 17, No. 1 (2006), Vol. 18 (2007), and Vol. 20 (2008). We wish to thank all of those who submitted papers and helped with the reviewing process. We also wish to thank all those who helped in organizing this workshop for their efforts in local arrangement, especially the Local Committee Members in Boston.

Scientific Committee
Charles DeLisi (Boston University, Chair)
Minoru Kanehisa (Kyoto University)
Edda Klipp (Humboldt-Universität zu Berlin)
Satoru Miyano (University of Tokyo)
Scott Mohr (Boston University)
Iwona Wallach (Charité - Universitätsmedizin Berlin)

SCIENTIFIC COMMITTEE

Charles DeLisi	Boston University, Chair
Minoru Kanehisa	Kyoto University
Edda Klipp	Humboldt-Universität zu Berlin
Satoru Miyano	University of Tokyo
Scott Mohr	Boston University
Iwona Wallach	Charité - Universitätsmedizin Berlin

STRUCTURAL FEATURES AND EVOLUTION OF PROTEIN-PROTEIN INTERACTIONS

JOACHIM VON EICHBORN

joachim.eichborn@charite.de

STEFAN GÜNTHER

stefan.guenther@charite.de

ROBERT PREISSNER

robert.preissner@charite.de

Structural Bioinformatics Group, Institute for Physiology, Charité–University Medicine, Arnimallee 22, 14195 Berlin, Germany

Solved structures of protein-protein complexes give fundamental insights into protein function and molecular recognition. Although the determination of protein-protein complexes is generally more difficult than solving individual proteins, the number of experimentally determined complexes increased conspicuously during the last decade. Here, the interfaces of 750 transient protein-protein interactions as well as 2,000 interactions between domains of the same protein chain (obligate interactions) were analyzed to obtain a better understanding of molecular recognition and to identify features applicable for protein binding site prediction.

Calculation of knowledge-based potentials showed a preference of contacts between amino acids having complementary physicochemical properties. The analysis of amino acid conservation of the entire interface area showed a weak but significant tendency to a higher evolutionary conservation of protein binding sites compared to surface areas that are permanently exposed to solvent. Remarkably, contact frequencies between outstandingly conserved residues are much higher than expected confirming the so-called "hot spot" theory. The comparisons between obligate and transient domain contacts reveal differences and point out that structural diversification and molecular recognition of protein-protein interactions are subjected to other evolutionary aspects than obligate domain-domain interactions.

Keywords: protein interactions; domain interactions; conservation; interaction hot spots.

1. Introduction

Proteins interact quickly and specifically with other biomolecules, especially other proteins [18]. Compared to the genomic data of complex organism that increased explosively during the last decades, knowledge about the human interactome is still very fragmentary [21]. The analysis of solved structures of protein-protein complexes give fundamental insights into molecular recognition and may help to obtain protein properties that can be used to predict specific interaction partners by *in silico* methods. The properties of protein-protein interaction sites have been studied since the 1970s [1, 3, 4, 12, 13] and results were integrated in some very useful tools for protein-protein docking [5, 7, 10]. Such tools can provide evidence for the right binding mode of two proteins. Docking is usually achieved by maximizing the shape and physicochemical complementarity through generation of large sets of possible conformations. However, sampling of relevant conformations and the discrimination of native-like configurations from the large number of non-native alternatives remain challenging.

To determine the relevance of a certain protein-protein interaction model (decoy), the residue contacts between proteins in the model can be evaluated. In 1995 a pioneering work of Clackson and Wells was published that has shown that just a small and complementary set of cooperative contact residues, termed "hot spots" maintains the largest part of the binding affinity [6]. Clackson and Wells distinguished between functionally important residues and surrounding residues that are less important. However, their analysis is based on a single complex structure of the human growth hormone and its receptor. In 2007 a similar line of reasoning was taken up by Shulman-Peleg et al. [20] based on a more extensive data set of 71 complex structures. They came to similar results, claiming that a few highly conserved residues are crucial for protein interactions.

Physical interactions between proteins are mostly controlled by their domains, the fundamental units in the evolution of genes and functions. Beside domain-domain contacts that develop temporarily during specific protein-protein interactions (transient interactions) cristallographic information on numerous domain-domain contacts between domains of the same multidomain protein are available in the Protein Data Bank (PDB) [2]. Such obligate interactions would provide a valuable resource for analysing molecular recognition, if they show similar properties of molecular recognition as transient interactions. An existing analysis addressing this question was performed 2005 and is based on a dataset consisting of 212 transient and 115 obligate protein complexes [16]. They come to the conclusion that some differences in evolutionary pressure exist between the two types of interactions. Obligate complexes evolve at a relatively slower rate, allowing them to coevolve with their interacting partners. Nevertheless the analysis is based only on selected domain family-family pairs representing only a small part of the structural information available today.

Here, we have analyzed characteristics of protein-protein interaction sites with the aim of describing them and to obtain properties that are suitable for improving protein docking. For this purpose we have performed a large–scale analysis based on a data set of 2,000 obligate and 750 transient domain-domain interfaces. Differences between surface regions and interaction sites as well as differences between obligate and transient interactions are pointed out. The hot spot hypothesis is rechecked and specific contacts between residues are systematically evaluated. Knowledge-based residue contact maps are provided for evaluation of protein-protein decoys and protein models.

2. Methods

2.1. *Data Set*

For this study the protein interaction sites provided by the JAIL-database [11] are used. JAIL contains an exhaustive set of 190,000 protein interaction sites. They are computed using the protein structures available from the PDB.

The data sets used for this study are based on all interactions between SCOP-defined protein domains [17] that are part of the JAIL database in the version from June 2009. We constructed two sets of interactions. The first one contains interactions between domains that are part of different proteins (transient). In contrast, the interactions in the second set have to be between domains on one amino amino acid chain (obligate).

Afterwards, both sets were filtered, so that interactions between domains of any two different SCOP-classes are represented by at most one interaction. Thereby, nonredundant sets of protein class interactions are guaranteed.

After filtering we obtained 750 transient interactions and 2,000 obligate ones.

2.2. *Interface Definition*

The residues that participate in a protein interaction are referred to as the "interface" in the following text. We derived the interface definition used by JAIL. According to this definition a complete residue is part of an interface if at least one atom of the amino acid is located within a range of 4.5 Å to any atom of the interacting domain. These residues are considered to be in contact with each other (Fig. 1). Additionally, one part of an interface must consist of at least 5 Cα atoms.

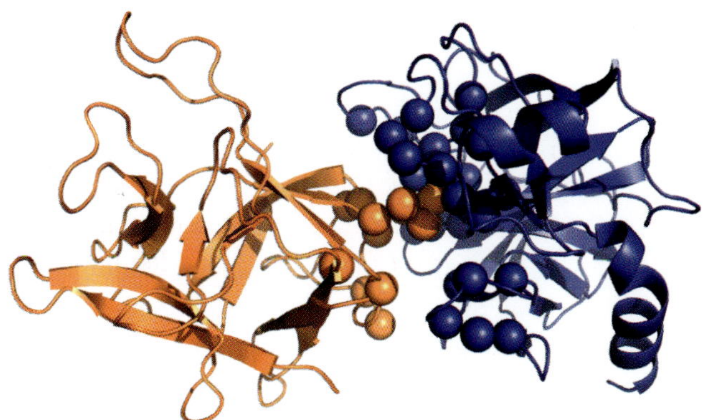

Fig. 1: Interface between two SCOP-domains (SCOP-Ids d1avxa_ and d1avxb_). The two interacting domains are shown in orange and blue. Cα atoms of residues that are in contact with residues of the other domain are shown as spheres.

2.3. *Surface Definition*

To determine the residues on the protein surface, we used calc-surface [8]. If only one atom of a residue is on the surface, the whole residue is considered to be part of the surface.

2.4. *Contact Frequency Normalization*

The preferences and aversions towards residue contacts are given as multiples of the expected values to correct for different frequencies of the participating amino acids. These multiples are computed by equation 1. x and y denote the interacting amino acids and $C(x, y)$ the number of contacts between them. $count(i)$ is the number of occurences of an amino acid i in interfaces and $count(contacts)$ is the total number of contacts that exist.

$$M(x, y) = \frac{C(x, y)}{\frac{count(x)}{\sum_{z \in AA}^{20} count(z)} \cdot \frac{count(y)}{\sum_{z \in AA}^{20} count(z)} \cdot count(contacts)} \tag{1}$$

The normalization of the contact values between residues of specific conservation grades is done analogously.

2.5. *Conservation Score*

The conservation information is taken from the ConSurf-DB [9]. The ConSurf-DB contains precomputed conservation scores for all structures in the PDB and is updated on a monthly basis. The conservation scores listed in the ConSurf-DB are computed using the Rate4Side algorithm [19] based on a multiple sequence alignment of homologue sequences.

Positive scores indicate variable positions and negative scores conserved ones. Afterwards, the scores are normalized, such that the average over all residues is zero and the standard deviation is one. Finally the scores are binned in the integer range of one to nine. These binned values are used in this study.

3. Results

We investigated the frequencies of the different amino acids in interfaces and on surfaces. Fig. 2 displays the results for the transient interfaces, the results for obligate interfaces are almost similar (not shown). The most remarkable difference between the groups is, that Glycine is nearly as frequent in obligate interfaces as it is on the surfaces whereas for transient interfaces Glycine is more frequent on the surfaces.

Fig. 3 shows the amino acids' interface/surface preferences as ratios between the interface and surface frequencies from Fig. 2. A characteristic pattern of interface/surface preferences can be observed. For example Cysteine, Tyrosine,

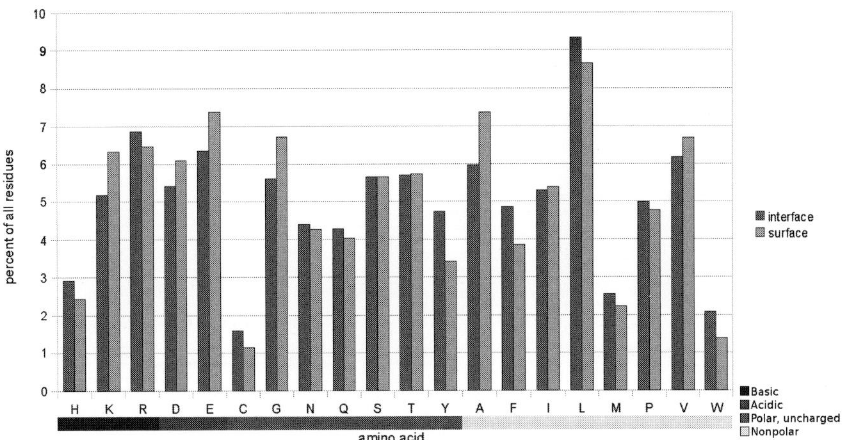

Fig. 2: Frequencies of the different amino acids in the interfaces and surfaces of transient interactions.

Phenylalanine and Tryptophan are considerably more frequent in interfaces than in surfaces.

In the following we examined the pairwise contact preferences of the different amino acid types. In the data sets used for this study each interface contains on average nearly 60 residue contacts. We created heat maps showing the preferences of all possible amino acid pairs to be in contact. To compute these values at first the absolute number of occurrences is counted for each amino acid pair. Those numbers are affected by the frequencies of the amino acids in the interfaces (the "interface" bars from Fig. 2). For example there will be a large number of contacts between two highly abundant amino acids, even if they have no preference for contact with each other. To deal with this, the observed number of contacts for each amino acid

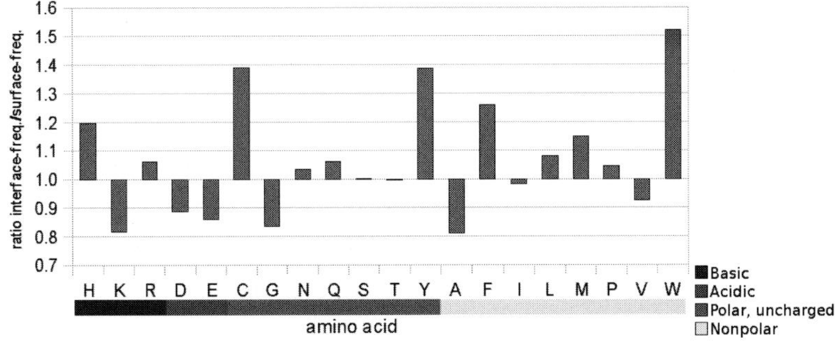

Fig. 3: Ratio of interface-frequency and surface-frequency for the different amino acids in transient interactions. Values >1 indicate interface preferences, values <1 surface preferences.

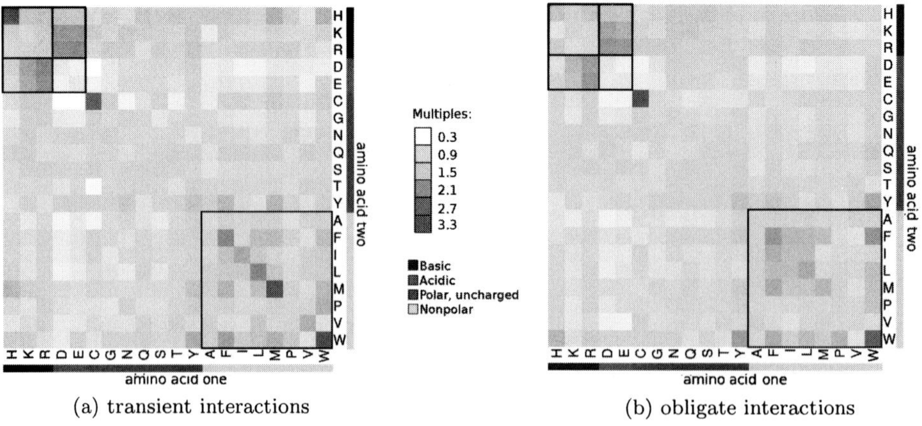

(a) transient interactions   (b) obligate interactions

Fig. 4: Multiples of the expected values that were observed for all amino acid contacts. The amino acids are grouped according to their physicochemical properties, some groups are highlighted by black rectangles.

pair is normalized with the expected number of contacts between the participating amino acids as described in the methods section.

We created these heat maps with normalized values for both kinds of interfaces (Fig. 4). The amino acids in the figures are grouped into basic, acidic, polar uncharged and nonpolar based on their physicochemical properties. In both maps contact preferences between some of these groups can be observed. The most striking examples are the groups of acidic and basic amino acids. Amino acids in these groups are only rarely in contact with other amino acids of the same group. However, they show a strong tendency to come into contact with amino acids of the other group. The only exception in these groups is Histidin, which does not prefer contacts to acidic amino acids, and shows a strong tendency to be in contact with itself in transient interfaces.

Another clear tendency is, that nonpolar amino acids are preferably in contact with other nonpolar amino acids.

As the contacts in the interfaces are crucial, an effect on the conservation of interface residues can be expected. Residue contacts in interfaces take place in close spatial proximity and preferably between residues that fit well together according to their physicochemical properties, as shown above. Because of these two factors the mutation rate can be expected to be lower in interfaces than on the surface.

As a matter of fact interfaces tend to be on average more conserved than surfaces. The average conservation of interfaces was 5.6 and 5.8 in the two interface groups. In contrast the average conservation of the surfaces was 5.2 and 5.1. So the overal conservation is higher in interfaces.

By looking at the ratio of the interface- and surface-frequencies of differentially conserved residues, the difference becomes even clearer (Fig. 5). Residues that are

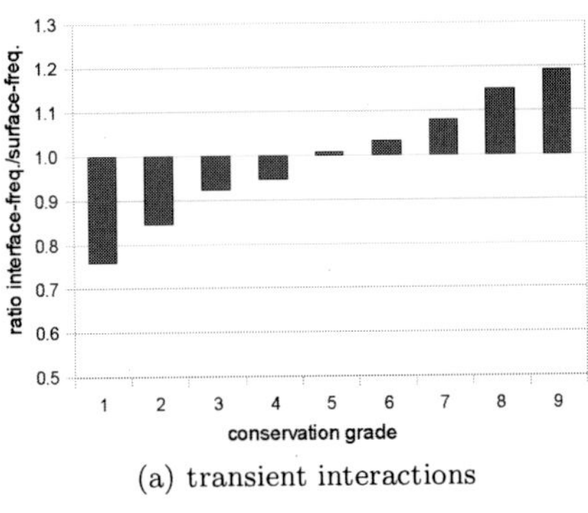
(a) transient interactions

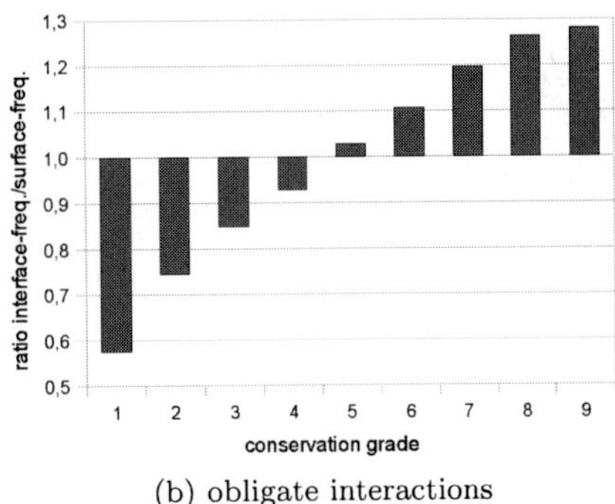
(b) obligate interactions

Fig. 5: Ratio of interface-frequency and surface-frequency for the residues of each conservation grade. Values >1 indicate interface preferences, values <1 surface preferences.

very variable (conservation grade 1) are more frequent in the surface than in the interface. In contrast, highly conserved residues (conservation grade 9) are more frequent in the interface. For the conservation grades between these extreme values the preference is shifting from surface to interface with rising conservation grade. The surface/interface preferences are stronger in obligate interfaces than in transient ones.

We analyzed, how the contact preferences of residues are affected by their conservation. In Fig. 6 the observed numbers of all possible combinations between residues of specific conservation grades are shown as multiples of the expected values. There is a clear tendency for residues, to be in contact with other residues that share the same conservation grade. Generally, a contact between residues becomes more and more unlikely with growing difference in their conservation grade.

Additionally there are peaks where very conserved and very variable residues interact with equally conserved/variable residues. The peak for very conserved residues interacting with each other supports previous findings that certain regions inside interfaces are crucial for interaction. The described pattern is even stronger for obligate interfaces than for transient ones.

4. Discussion

The described results show, that physicochemical properties of amino acids have a considerable influence on their contact preferences in interfaces. When amino acids are grouped by their physicochemical properties, a general pattern of preferences and aversions for contacts between members of the different groups can be observed. Additionally, there are some remarkable contacts between single amino acid pairs. The most strinking example is Cysteine-Cysteine, which shows the highest preference of all contacts. We guess the reason for this is, that Cysteines are able to form disulfide bonds with each other. These bonds can help to strengthen reversible and stable interactions. However, Cystein-Cystein contacts should be more frequent in obligate interfaces than in transient interfaces on the surfaces. Indeed there are 2.6 times more Cystein-Cysteine contacts than expected in the transient interfaces. In obligate interfaces this ratio rises to 3.5.

Another set of amino acids, that show a remarkable behavior, are the aromatic amino acids Tyrosine, Phenylalanine and Tryptophan. These amino acids show strong preferences to be in contact with other amino acids from the set. A plausible explanation for this is, that aromatic amino acids can interact via aromatic stacking. This may also be the reason for the high frequency of Histidin-Hisitdin contacts on transient interfaces.

The second focus of this work was to look for relations between the contacts in interfaces and evolutionary constraints. We observed, that conserved residues are more likely to be in an interface than on the rest of the surface. This preference gets weaker with decreasing conservation and turns into a surface preference for variable residues.

This observation is congruent with previous findings [15]. Interfaces are regions where the participating domains or proteins interact with each other specificly and selectively. Therefore, the interaction partners have to be able to identify each other by having complementary interaction sites. To achieve high levels of selectivity and specificity, the interacting sites have to fit together well. This enforces a high conservation of interaction sites as mutations may inhibit the binding of interaction partners.

The preferences of variable residues for the surface and of conserved ones for the interface are stronger in transient interfaces than in obligate ones. This supports previous findings by Mintseris and Weng [16], that obligate interfaces evolve slower than transient ones. Mintseris and Weng argument, that obligate interactions are

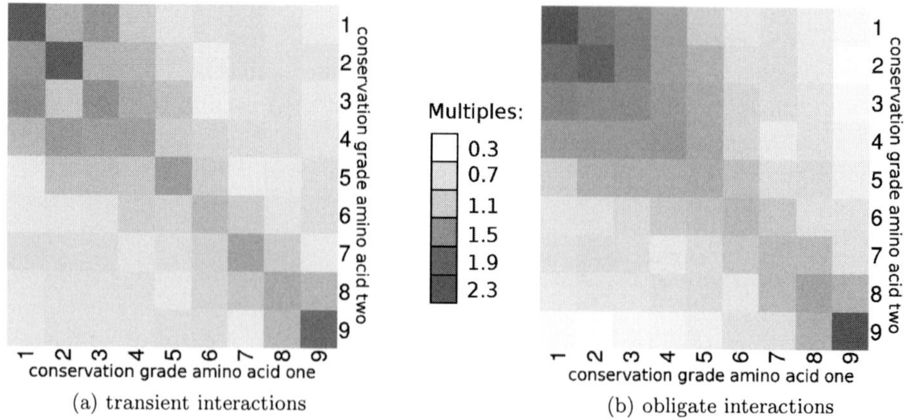

(a) transient interactions   (b) obligate interactions

Fig. 6: Multiples of the expected values that were observed for contacts between residues of specific conservation grades.

less frequently mutating, giving the interaction partner the chance to coevolve. On the opposite transient interfaces have to be able to adapt to changes of the interaction partner rather fast, leading to a lower conservation.

Additionally, we examined, whether the residues' conservations are related to their probability of being in contact with each other. There is a very clear picture, that residues prefer to be in contact with other residues that have the same grade of conservation. That means, that conserved residues are not randomly distributed on both sides of the interface. Very variable and very conserved residues especially prefer to interact with equally conserved residues. This finding supports the theory, that a few residues in an interface are crucial for interaction [14]. These residues have to be present on both sides of the interface, therefore they have to be strictly conserved in order to maintain the interaction. On the other hand, very variable residues also show a strong preference for contact with each other. The argument for this follows from the one above. If a residue-residue contact is important, both participating residues have to be conserved. If one residue of a contact ist very variable, the contact seems to be rather unimportant for the interaction. Therefore the other residue can be variable, too.

This also explains why residues, whose interaction grades are very different only interect about half as often as expected. This combination is not reasonable as the conservation of just one partner does not ensure that the interaction will always be working.

5. Conclusion

We presented properties that can help to identify protein binding sites on the surface of proteins. This information can be used to locate functionally important structural polymorphisms or to determine the near native model in a set of protein docking decoys. Additionally, characteristics of the residue contacts between domains were systematically analyzed and are shown as contact maps. They may be valuable for further refinement of protein-protein complex models.

Although obligate interactions are more conserved than transient interactions, they share most of their physicochemical properties. Therefore, obligate interactions can be used as models for protein-protein interactions.

Acknowledgements

This work was supported by the International Research Training Group Boston-Kyoto-Berlin, funded by the German Research Foundation (DFG).

References

[1] Argos, P., An investigation of protein subunit and domain interfaces, *Protein Eng.*, 2(2):101–113, 1988.

[2] Berman, H. M., Westbrook, J., Feng, Z., Gilliland, G., Bhat, T. N., Weissig, H., Shindyalov, I. N., Bourne, P. E., The protein data bank, *Nucleic Acids Res.*, 28(1):235–242, 2000.

[3] Chothia, C., Janin, J., Principles of protein-protein recognition, *Nature*, 256(5520):705–708, 1975.

[4] Chakrabarti, P., Janin, J., Dissecting protein-protein recognition sites, *Proteins*, 47(3):334–343, 2002.

[5] Chen, R., Li, L., Weng, Z., Zdock: an initial-stage protein-docking algorithm, *Proteins*, 52(1):80–87, 2003.

[6] Clackson, T., Wells, J. A., A hot spot of binding energy in a hormone-receptor interface, *Science*, 267:383–386, 1995.

[7] Comeau, S. R., Gatchell, D. W., Vajda, S., Camacho, C. J., Cluspro: a fully automated algorithm for protein-protein docking, *Nucleic Acids Res.*, 32(Web Server issue):W96–99, 2004.

[8] Gerstein, M., Tsai, J., Levitt, M., The volume of atoms on the protein surface: calculated from simulation, using voronoi polyhedra, *J. Mol. Biol.*, 249(5):955–966, 1995.

[9] Goldenberg, O., Erez, E., Nimrod, G., Ben-Tal, N., The consurf-db: pre-calculated evolutionary conservation profiles of protein structures, *Nucleic Acids Res.*, 37(Database issue):D323–327, 2009.

[10] Günther, S., May, P., Hoppe, A., Frömmel, C., Preissner, R., Docking without docking: Isearch–prediction of interactions using known interfaces. *Proteins*, 69(4):839–844, 2007.

[11] Günther, S., von Eichborn, J., May, P., Preissner, R., Jail: a structure-based interface library for macromolecules, *Nucleic Acids Res.*, 37(Database issue):D338–341, 2009.

[12] Hubbard, S. J., Argos, P., Cavities and packing at protein interfaces, *Protein Sci.*, 3(12):2194–2206, 1994.

[13] Jones, S., Thornton, J. M., Analysis of protein-protein interaction sites using surface patches, *J. Mol. Biol.*, 272(1):121–132, 1997.

[14] Ozlem Keskin, O., Ma, B., Nussinov, R., Hot regions in protein–protein interactions: the organization and contribution of structurally conserved hot spot residues, *J. Mol. Biol.*, 345(5):1281–1294, 2005.

[15] Ma, B., Elkayam, T., Wolfson, H., Nussinov, R., Protein-protein interactions: structurally conserved residues distinguish between binding sites and exposed protein surfaces, *Proc. Natl. Acad. Sci. U.S.A.*, 100(10):5772–5777, 2003.

[16] Mintseris, J., Structure, function, and evolution of transient and obligate protein-protein interactions, *Proceedings of the National Academy of Sciences*, 102(31):10930–10935, 2005.

[17] Murzin, A. G., Brenner, S. E., Hubbard, T., Chothia, C., Scop: a structural classification of proteins database for the investigation of sequences and structures, *J. Mol. Biol.*, 247(4):536–540, 1995.

[18] Pawson, T., Nash, P., Assembly of cell regulatory systems through protein interaction domains, *Science*, 300(5618):445–452, 2003.

[19] Pupko, T., Bell, R. E., Mayrose, I., Glaser, F., Ben-Tal, N., Rate4site: an algorithmic tool for the identification of functional regions in proteins by surface mapping of evolutionary determinants within their homologues, *Bioinformatics*, 18 Suppl 1:S71–77, 2002.

[20] Shulman-Peleg, A., Shatsky, M., Nussinov, R., Wolfson, H. J., Spatial chemical conservation of hot spot interactions in protein-protein complexes, *BMC Biol.*, 5:43, 2007.

[21] Stumpf, M. P. H., From the cover: Estimating the size of the human interactome, *Proceedings of the National Academy of Sciences*, 105(19):6959–6964, 2008.

GRAPHICAL ANALYSIS AND EXPERIMENTAL EVALUATION OF *SACCHAROMYCES CEREVISIAE* $P_{TRK_{1|2}}$ AND $P_{BMH_{1|2}}$ PROMOTER REGION

SUSANNE GERBER[1]
gerber@molgen.mpg.de

GUIDO HASENBRINK[2]
uzs6br@uni-bonn.de

WOUTER HENDRIKSEN[3]
w.t.hendriksen@biology.leidenuniv.nl

PAUL VAN HEUSDEN [3]
G.P.H.van.Heusden@biology.leidenuniv.nl

JOST LUDWIG[2]
jost.ludwig@uni-bonn.de

EDDA KLIPP [1]
edda.klipp@rz.hu-berlin.de

HELLA LICHTENBERG-FRATÉ[2]
h.lichtenberg@uni-bonn.de

[1] *Group of Theoretical Biophysics, MNF I,Humboldt University of Berlin, Invalidenstrasse 42, 10115 Berlin, Germany*
[2] *University of Bonn, IZMB, Kirschallee 1 - D 53115 Bonn ,Germany*
[3] *Section Molecular and Developmental Genetics, Institute of Biology, Leiden University,Sylviusweg 72, 2333 BE Leiden, The Netherlands*

We designed a simple graphical presentation for the results of a transcription factor (TF) pattern matching analysis. The TF analysis algorithm utilized known sequence signature motifs from several databases. The graphical presentation enabled a quick overview of potential TF binding sites, their frequency and spacing on both DNA strands and thus straight forward identification of promising candidates for further experimental investigations. The developed tool was applied on in total four *Saccharomyces cerevisiae* gene promoter regions. The selected differentially expressed genes belong to functionally different families and encode duplicate functions, TRK1 and TRK2 as ion transporters and BMH1 and BMH2 as multiple regulators. Output evaluation revealed a number of TFs with promising differences in the promoter regions of each gene pair. Experimental investigations were performed by using corresponding TF yeast mutants for either phenotypic analysis of ion transport mediated growth or expression analysis of BMH1,2 genes. Upon phenotypic testing one TF mutant exhibited severely impaired growth under non-permissive conditions. This TF, Mot3p was identified as of most abundant potential binding sites and distinctive patterns among the TRK promoter regions.

Keywords: *Saccharomyces cerevisiae*; promoter region; graphical analysis tool.

1. Introduction

In the post-genome era the task of regulatory elements prediction and in particular promotor region analysis has received increasing interest and a considerable number of algorithms, tools and analysing strategies was developed. However, despite many useful approaches signal finding such as pattern discovery in DNA sequences is still a fundamental problem in molecular biology with such important applications as locating regulatory sites and drug target identification. Most signals in DNA sequences are indeed complex or not yet discovered, such that there is a lack of good models or reliable algorithms for their authentic recognition. For instance, some transcription factor binding sequences are injective but the majority show a rather high degree of wobbling which means in mathematical terms they show a bijective behaviour. Concerning the model organism *S. cerevisiae* the currently available tools like AlignACE [10], Consensus [6], TESS [16, 27], TRANSFAC [23] or Genomatix [25] differ from each other basically in their definitions of what constitutes a motif, what constitutes statistical overrepresentation of a motif and the methods used to identify statistically overrepresented motifs (for a review see [18]). All these tools differ enormous in their binding site predictions or recognize motifs that are not even specific for the queried organism, e.g. mammalian TF motifs in the promoter regions of *S.cerevisiae*. In all cases the graphical presentation of positive hits on both strands is a highly condensed and information-wise insufficient image. An additional obstacle for wet lab scientists might be the fact that they are offered little guidance in the choice among these tools. Since there have been published only a few assessments or reviews of some of these motif discovery tools ([15] and [18]) and it is virtually impossible to decide on objective grounds of whether the results of the tool (A) are better or more plausible than those of tool (B). In conjunction with a study focusing on cation homeostasis, ion transporters and their potential regulators in yeast we collated a dataset of TFs (sequence signature motifs) known as functional in *S. cerevisiae*. The challenge was to provide on the basis of this comprehensive dataset a bioinformatical analysis that should reveal first indications of potential TFs for specific genes, or their upstream regulating sequences respectively, involved in cation transport or their possible regulators. The analysis emphasized on frequency, spacing and accuracy of match and presented the results in a graphical format enabling straight forward visual inspection and decisions on potentially promising candidates for further experimental investigation.

As examples for an explorative approach the promoter regions of the TRK1 and TRK2 and BMH1 and BMH2 were selected. These genes encode duplicated functions in yeast, are differentially expressed and highly regulated and appeared thus as reasonable candidates for a comparative promoter region analysis. The first ones are components of the potassium transport system [3, 11, 12, 24], and latter, belonging to the group of 14-3-3 proteins, act as multiple regulators. 14-3-3 proteins are a family of highly conserved proteins and have been identified in all eukaryotic species investigated. *S. cerevisiae* has two genes encoding 14-3-3 proteins, BMH1 (major isoform) and BMH2, and these proteins are involved in various cellular processes [21]. Numerous studies revealed that expression of BMH1 and BMH2 is affected by different circumstances (for a review see [22]). Expression of BMH1 and BMH2 might therefore be tightly regulated by transcription factors.

2. Material and Methods

2.1. *Bioinformatical Analysis*

We collected a data set of all available TFs known to be functional in yeast within the literature documented binding sites. For collection, we parsed the YEASTRACT database [17, 28] (Yeast Search for Transcriptional Regulators And Consensus Tracking) that presently contains 30990 regulatory associations between the yeast genes based on more than 1000 bibliographic references [17]. Each regulation has been manually annotated after examination of the relevant references. Further information about each yeast gene was obtained from the Saccharomyces Genome Database (SGD) [9], Regulatory Sequence Analysis Tools (RSAT) [20] and Gene Ontology (GO) Consortium [19]. The assembled dataset for this study contained the description of 288 specific and in the literature documented DNA binding sites for a group of 108 transcription factors.

The analysis was performed by implementing a Matlab [26] program that imports a promotor sequence in fasta-format as well as the data set of TFs with their dedicated potential binding sequences. If the TF binding sequence contains IUPAC codes other than one of the four nucleotide the program will identify all possible matches. For example, the symbol $'R'$ in a sequence represents a purine at this position (either G or A) and the algorithm would accomplish a pattern search for each sequence on the promotor. For each match the start and stop position on the query promotor promoter sequence is stored, as well as the probability of a match of this sequence by just chance. If more than one match of the investigated sequence is identified, the distances between all possible binding sites of this TF are calculated. At the end of the analysis the user will receive a table with the information of every potential binding sequence combined to the following documentation: the names of the tested TFs and the related sequences; the average likelihood for every sequence to bind on this promoter just by chance; for succesfull matches the start and stop positions on both strands and, in case that more than one succesful match was found, the distances between all these potential binding sites. Furthermore - as a highlight - a graphical presentation of all potential binding sites on the forward and backward strands for every promotor region is a key add on.

2.2. *Yeast Strains, Media and Growth Tests*

Experiment 1

Haploid *S. cerevisiae* wild type, potassium transporter and TF mutant strains (BY4741; [2]; Saccharomyces Genome Deletion Project) were used throughout this study as listed in Table 1. Strains were grown aerobically at 28 °C with rotational shaking at 250 rpm in liquid amino acid, ammonium sulphate and potassium free YNB-F media (CYN7505, Formedium). The medium was supplemented with a complete supplement mixture (DBSM225, Formedium), ammonium sulphate, 2% D-glucose and, for precultures 50 mM KCl. For test cultures KCl was added to concentrations as indicated. All medium components were dissolved in ultrapure MilliQ water and pH was adjusted to 5.8 with ammonium hydroxide. Stationary phase cells were washed three times in MilliQ water and diluted in fresh YNB-F medium to a start optical density OD600 of 0.4 (pathlength 1 cm, Pharmacia Ultrospec 2000 Spectrophotometer, corresponding to 6x106 cells per mL) with respective KCl concentration. Growth assessment was obtained by photometric determination of the turbidity at 600 nm in 15-min intervals in transparent 96-well microtiter plates using a microplate reader (Tecan, Infinite F200) with 28 °C incubation and constant agitation for 17 h. Growth calculation and statistical analysis was performed by integral determination (approximation of the area under the growth curves) obtained within 17 h incubation according to [5]. Subsequently values for growth were normalized to those of wild type strain values at 50 mM KCl.

Table 1: Selected transcription factors for experimental analysis

| Transcription Factor | TRK1 | | | TRK2 | | |
matches	foreward	reverse	total	foreward	reverse	total
Fkh1p	4*	5*	9	3*	1	4
Fkh2p	4*	5*	9	3*	1	4
Mot3p	2*	11*	13	6	1	7
Msn2p		1	1	3		3
Msn4p		1	1	3		3
Stb5p	3	1	4	4	6	10

Note: * Total number of hits with more than one possible binding sequence

Experiment 2

In a second experimental scenario the *S. cerevisiae* BY4743 wild type *MATa/α his3Δ1/his3Δ1 leu2Δ0/leu2Δ0 LYS2/lys2Δ0 met15Δ0/MET15 ura3Δ0/ura3Δ0*) and transcription factor mutant (*Saccharomyces* Genome Deletion Project [4]) cells were grown exponentially in YPD broth to an OD_{620} of 0.6-0.8. The cells were harvested by immediately freezing with liquid nitrogen. Total RNA was isolated using the RiboPure^{TM}-Yeast Ambion RNA isolation kit (Ambion) according to the manual. This procedure included a DNase treatment to breakdown all chromosomal DNA. Subsequently, cDNA was produced using the Superscript III Reverse Transcriptase (Invitrogen), according to the manual. cDNA was diluted two times and 1 μl was used as a template for qPCR using the DyNAmo^{TM} Probe qPCR Kit (Finnzymes, Bioke). ACT1, BMH1, and BMH2 relative gene expression was measured on the Chromo4^{TM} Real-Time Detector (Biorad) using Opticon3 motion software (Biorad). The oligonucleotide primers used in this study are listed in Table 2. BMH1 and BMH2 gene expression ratios were calculated using the the $\Delta\Delta C(t)$ method [13], and these ratios were normalized to ACT1 expression and normalized to wild type BMH1 and BMH2 expression. This procedure was performed for two biological duplicates and expression was measured in triplicate and in duplicate for every sample.

Table 2: Oligonucleotide primers

Primer name	Sequence 5'-3'	Target
WH11	AGAGTTGCCCCAGAAGAACA	ACT1
WH12	GGCTTGGATGGAAACGTAGA	ACT1
WH29	CGTGCTAGACTCCCACTTAATT	BMH1
WH30	GGCCTTTTCTCTAGCATCG	BMH1
WH15	GTTAGATTCTCATTTAATCCCTTCT	BMH2
WH16	ATAAGCCTCCAAAGAGGAGTTG	BMH2

3. Results

The complete resulting output table for all 288 tested TF sequence motifs for genes TRK1,2 and for BMH1,2 are accessible in the supplementary material section, as well as the created graphics. As an example the graphical output of the TRK1 promotor analysis is given in Fig. 1. Out of the entire output list promising candidates for experimental investigations were selected along the following criteria: Sequence unambiguousness in the pattern matching of TF sequence signature to the DNA; the frequency of hits and significant differences in the frequency of hits between the analysed promoter regions of the gene pairs; a rational spacing between the potential TF binding sites (predicted potential binding sites that were at close proximity would constrain each other and were thus devaluated).

Consequently for the TRK1,2 genes 10 potential TFs were selected for further experimental analysis and for the Bmh1,2 genes 7 TFs. Since in the cation homeostasis project we are also interested in other cation transport systems of *S. cerevisiae* the promoter regions of the *ENA1, TOK1, PHO89, NHA1* and *PMA1* genes were subjected to computational TF sequence signature analysis including graphical presentation. However, these potentially promising transcription factor candidates have not yet been experimentally investigated. Calculations and graphics are provided in in supplementary material.

Experiment 1

In *S. cerevisiae* tight control and modulation of the intracellular K^+ concentration (cellular K^+ homeostasis) and corresponding ion fluxes is crucial for cell metabolism, surveillance and growth. Two closely related plasma membrane localised K^+ translocation systems, TRK1p and TRK2p mediate K^+ uptake to

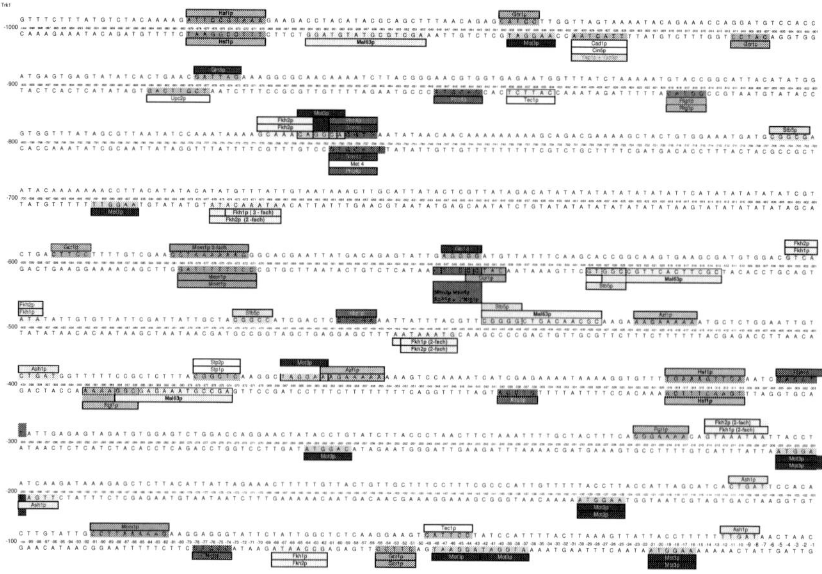

Fig. 1: Graphical analysis of TRK1 promotor regions.

maintain a high intracellular potassium concentration. *S. cerevisiae* strains devoid of TRK1 (ΔTRK1) or TRK1 and TRK2 (ΔTRK1,2) are defective in potassium uptake and exhibit impaired growth on limited external potassium con-centrations. From the potassium transporters TRK1 and TRK2 promoter region binding site analysis in total 10 potential TFs were obtained. Selection of these TFs or respective mutants as potentially important for further experimental analysis followed the simple rationale of clear frequency and pattern difference within the two promoter regions. Of these, four TF mutant strains are reported as not viable (Saccharomyces Genome Database, SGD). The methodological approach for the remaining six TF mutants was set accordingly: Will the lack of these TFs influence the TRK-system mediated potassium uptake detectable as growth differences under limited potassium conditions? The mutant yeast strains Δmot3, Δmsn2, Δmsn4, Δstb5, Δfkh1, Δfkh2 carry disruptions in the respective TF genes. The growth of these strains was compared to that of *S. cerevisiae* wild type, ΔTRK1 und ΔTRK1,2 under potassium limited and permissive conditions (Fig. 2). At 10 μM KCl$_{ext}$ none of the strains did grow. At 1 mM KCl$_{ext}$ considerable differences between the strains were observed. The ΔTRK1 strain, expressing only the TRK2 transporter grew only

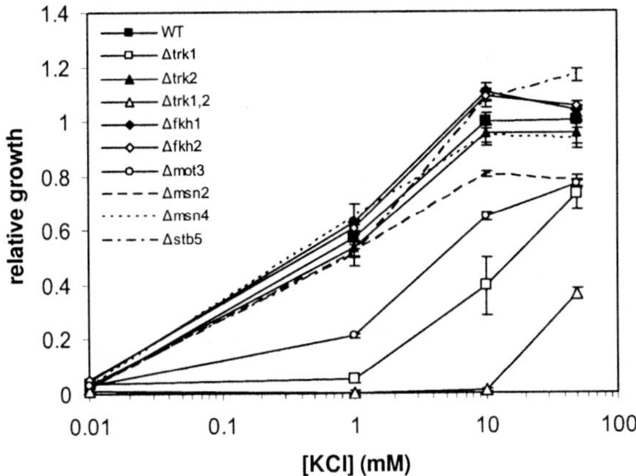

Fig. 2: Analysis of the survey data from experiment 1.

marginally above inoculum. In contrast, the ΔTRK2 strain, expressing solely TRK1 grew comparable to the wild type. The double ΔTRK1,2 mutant showed no growth at all. These results were consistent with previously published data [1, 5]. The strains with disruptions in the transcriptions factors FKH1/2, MSN2/4 und STB5 exhibited growth comparable to the wild type. However, growth of the TF mutant Δmot3 strain was considerably reduced compared to the wild type and the ΔTRK2 strain ($\sim$ 60%). At 10 mM KCl_{ext} the potassium transporter mutant ΔTRK1, ΔTRK1,2 and the Δmot3 strains showed with 51-64% and 20% respectively, reduced growth compared to the wild type. Mutant strains Δfkh1/2 and Δstb5 exhibited slightly enhanced growth ($\sim$ 10%). At 50 mM KCl_{ext}, ΔTRK1,2, ΔTRK1, Δmot3, Δmsn2 showed impaired growth of 64%, 27%, 24% and 22% compared to the wild type. Strains Δfk1/2 and Δstb5 showed similar growth as the wild type.

Experiment 2

Using mutants for TFs predicted to bind to one or both of the upstream regions of BMH1 and BMH2 (see suplementary material of the bioinformatical output), we tested the effect on expression during exponential growth in YPD medium. The used strains were null mutants of MSN2, RTG1, WAR1, RLM1, ASH1, FKH1, or GSM1. Grown in YPD medium, we did not observe any differences in growth of these strains. Compared to wild-type BY4743, BMH1 expression did not significantly change in the tested mutants and, hence, is probably not be regulated under the tested condition (Fig. 3A). The same was observed for BMH2 expression (Fig. 3B). We used the arbitrary twofold up- or downregulation as cut-off value for differential expression. However, subtle regulation by these transcription factors (i.e., fold change not higher than 2 and not lower than 0.5) cannot be excluded. In addition, other TFs may take over the regulatory role of the ones tested in this study. For instance, partial redundancy has been described for several TFs, e.g., MSN2/MSN4 and FKH1/FKH2 [8, 14]. In conclusion, we were not able to show an effect on regulation of BMH1 and BMH2 in the TFs mutants tested in this study.

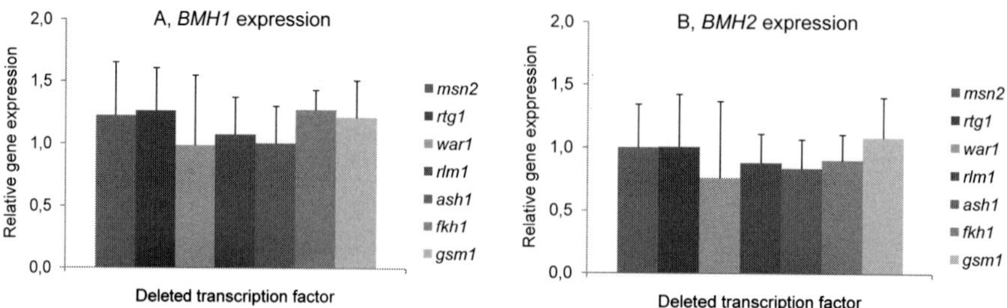

Fig. 3: Relative gene expression of BMH1 (A) and BMH2 (B) in BY4743 Δ/Δ msn2, Δ/Δrtg1, Δ/Δwar1, Δ/Δrlm1, Δ/Δash1, Δ/Δfkh1, Δ/Δgsm1 measured by real-time PCR. Gene expression was normalized to ACT1 expression and expression of BMH1 and BMH2 in BY4743 wild-type.

4. Discussion

Experiment 1

Six potential transcription factors with differential frequency and binding site patterns in the potassium transporters TRK1 and TRK2 promoter regions were subjected to growth tests of respective yeast mutant strains for potassium dependent growth. Disruption of the Fkh1/2 TFs (9 potential binding sites in TRK1 and 4 in TRK2 promoter region) caused at limiting potassium conditions no reduction in mitotic growth. Both TFs are important in the cell cycle, FKH1 as negative regulator and Fkh2 as positive regulator within the elongation process during transcription. Deletion of the negative regulator caused slight enhancement of growth under permissive conditions. For the Msn2/4p stress transcription factors (ratio TRK1:TRK2 of 1: 3) on TRK1 region only one potential site on the reverse strand was found whereas on the TRK2 region three potential sites were found on the forward strand. Deletion of these stress TFs had minor consequences on the K^+ growth phenotype. We conclude that Msn2/4 TFs are inferior for the upregulation of TKR transporters under K^+ stress conditions. Deletion of Stb5 (4 potential binding sites in TRK1 and 10 in TRK2 promoter region) was, according to the distribution pattern expected to impact mainly on the low affinity TRK2 expression. The Δstb5 mutant did grow slower than a ΔTRK2 disruption (entire K^+ uptake via TRK1), indicating that even a reduced (up) regulation of TRK2 has, with a present TRK1, only a minor impact. The Δmot3 strain exhibited a growth defect under K^+ limiting conditions of approximately half the effect of disruption of the potassium transporter Trk1p. Mot3p is known as involved in repression of several

DAN/TIR genes during aerobic growth, and repression of ergosterol biosynthetic genes. Under the assumption of a direct influence of Mot3p the lack of this TF could be responsible for a reduced expression of TRK1 (13 potential binding sites in TRK1 versus 7 in TRK2). However, alternatively, given the described role of Mot3p in ergosterol biosynthesis pathway, the deletion of this TF might pose an altered -for K^+ uptake transporters unfavourable- lipid composition onto the plasma membrane. In summary, the results from these initial phenotypic experiments are indeed interesting and hold the potential for extension towards construction and analysis of yeast strains bearing combinations of TF and potassium transporter mutations.

Experiment 2

No significant differences were found for BMH1 and BMH2 gene expression in the tested TFs mutants However, a regulation cannot be excluded. In addition, other TFs may take over regulatory role of the ones tested in this study. Another possibility is that the tested TFs are not active during exponential growth in a rich medium like YPD. In conclusion, we were not able to show an effect on regulation of BMH1 and BMH2 in the TF mutants tested in this study.

5. Acknowledgments

This work was supported by the German Ministry for Eduction and Research (BMBF grant for the SysMO Eranet project Translucent) and by the German Research Foundation (DFG grant for the International Research Training Group "Genomics and Systems Biology of Molecular Networks").

References

[1] Bertl, A., et al., Characterization of potassium transport in wild-type and isogenic yeast strains carrying all combinations of TRK1, TRK2 and TOK1 null mutations, *Mol Microbiol*, 47(3):767–780, 2003.

[2] Brachmann, C.B., Davies, A., Cost, G.J., Caputo, E., Li, J., Hieter, P., Boeke, J.D., Designer deletion strains derived from *Saccharomyces cerevisiae* S288C: a useful set of strains and plasmids for PCR-gene disruption and other applications, *Yeast*, 14:115–132, 1998.

[3] Gaber, R.F., et al. TRK1 encodes a plasma membrane protein required for high-affinity potassium transport in Saccharomyces cerevisiae, *Mol Cell Biol*, 8(7):2848–2859, 1988.

[4] Giaever, G., et al., Functional profiling of the *Saccharomyces cerevisiae* genome., *Nature*, 25(418)(6896):387–391,2002

[5] Hasenbrink, G., Schwarzer, S., Kolacna, L., Ludwig, J., Sychrova, H., Lichtenberg-Fraté, H., Analysis of the mKir2.1 channel activity in potassium influx defective strains determined as changes in growth characteristics, *FEBS Letters*, 579(7):1723–1731, 2005.

[6] Hertz, G.Z., Stormo, G,D., Identifying DNA and protein patterns with statistically significant alignments of multiple sequences, *Bioinformatics*, 15(7-8):563–577, 1999.

[7] Hobby, J.D., *A User's Manual for MetaPost*, Tech. Rep. 162, AT&T Bell Laboratories, Murray Hill, New Jersey, 1992.

[8] Hollenhorst, P.C., Bose, M.E., Mielke, M.R., Mller, U., Fox, C.A., Forkhead genes in transcriptional silencing, cell morphology and the cell cycle. Overlapping and distinct functions for FKH1 and FKH2 in *Saccharomyces cerevisiae, Genetics*, 154(4):1533–1548, 2000.

[9] Hong, E.L., Balakrishnan, R., Christie, K.R., Costanzo, M.C., Dwight, S.S., Engel, S.R., Fisk, D.G., Hirschman, J.E., Livstone, M.S., et al., Saccharomyces Genome Database, http://www.yeastgenome.org/, 2007.

[10] Hughes, J.D., Estep, P.W., Tavazoie, S., Church, G.M., Computational identification of cis-regulatory elements associated with groups of functionally related genes in *Saccharomyces cerevisiae, J Mol Biol.*, 296(5):1205–1214, 2000.

[11] Ko, C.H., et al., TRK2 is required for low affinity K^+ transport in *Saccharomyces cerevisiae, Genetics*, 125(2):305–312, 1990.

[12] Ko, C.H., Gaber, R.F., TRK1 and TRK2 encode structurally related K^+ transporters in *Saccharomyces cerevisiae, Mol Cell Biol*, 11(8):4266–4273, 1991.

[13] Livak, K.J., Schmittgen, T.D., Analysis of relative gene expression data using real-time quantitative PCR and the 2(-Delta Delta C(T)) Method, *Methods*, 2(4):402–408, 2001.

[14] Martinez-Pastor, M.T., et al., The *Saccharomyces cerevisiae* zinc finger proteins Msn2p and Msn4p are required for transcriptional induction through the stress response element (STRE), *EMBO J* 15(9):2227–2235, 1996.

[15] Pevzner, P.A., Sze, S.H., Combinatorial approaches to finding subtle signals in DNA sequences, *Proc Int Conf Intell Syst Mol Biol.*, 8:269–278, 2000.

[16] Schug, J., Overton, G.C., *TESS: Transcription Element Search Software on the WWW*, Tech. Rep. CBIL-TR-1997-1001-v0.0, Computational Biology and Informatics Laboratory, School of Medicine, University of Pennsylvania, 1997.

[17] Teixeira, M.C., Monteiro, P., Jain, P., Tenreiro, S., Fernandes, A.R., Mira, N.P., Alenquer, M., Freitas, A.T., Oliveira, A.L., et al., The YEASTRACT database: a tool for the analysis of transcriptional relatory associations in *Saccharomyces cerevisiae, Nucleic Acids Res.*, 34 (Database Issue), D446–D451, 2006.

[18] Tompa, M. et al, Assessing computational tools for the discovery of transcription factor binding sites, *Nature Biotechnology*, 23: 137–144, 2005.

[19] The Gene Ontology Consortium, Gene Ontology: tool for the unification of biology, *Nat. Genet.*, 25(1): 25-29, 2000.

[20] van Helden, J., Regulatory sequence analysis tools, *Nucleic Acids Res.*, 31:3593–3596, 2003.

[21] van Heusden, G.P., Steensma, H.Y., Yeast 14-3-3 proteins, *Yeast*, 23(3):159–171, 2006.

[22] van Heusden, G.P., 14-3-3 proteins: Insights from genome-wide studies in yeast, *Genomics*, [Epub ahead of print], 2009.

[23] Wingender, E., Chen, X., Fricke, E., Geffers, R., Hehl, R., Liebich, I., Krull, M., Matys, V., Michael, H., et al., The TRANSFAC system on gene expression regulation, *Nucleic Acids Res.*, 29:281–283, 2001.

[24] Yenush, L., et al., pH-Responsive, posttranslational regulation of the Trk1 potassium transporter by the type 1-related Ppz1 phosphatase, *Mol Cell Biol*, 25(19):8683-8692, 2005.

[25] http://www.genomatix.de/

[26] http://www.mathworks.com/

[27] http://www.cbil.upenn.edu/tess

[28] http://www.yeastract.com/

STRATEGIES OF NON-SEQUENTIAL PROTEIN STRUCTURE ALIGNMENTS

AYSAM GUERLER

guerler@chemie.fu-berlin.de

ERNST-WALTER KNAPP

knapp@chemie.fu-berlin.de

Freie Universität Berlin, Institute of Chemistry and Biochemistry, Fabeckstrasse 36a, 14195 Berlin, Germany

Due to the large number of available protein structure alignment algorithms, a lot of effort has been made to define robust measures to evaluate their performances and the quality of generated alignments. Most quality measures involve the number of aligned residues and the RMSD. In this work, we analyze how these two properties are influenced by different residue assignment strategies as employed in common non-sequential structure alignment algorithms. Therefore, we implemented different residue assignment strategies into our non-sequential structure alignment algorithm GANGSTA+. We compared the resulting numbers of aligned residues and RMSDs for each residue assignment strategy and different alignment algorithms on a benchmark set of circular-permuted protein pairs. Unfortunately, differences in the residue assignment strategies are often ignored when comparing the performances of different algorithms. However, our results clearly show that this may strongly bias the observations. Bringing residue assignment strategies in line can explain observed performance differences between entirely different alignment algorithms. Our results suggest that performance comparison of non-sequential protein structure alignment algorithms should be based on the same residue assignment strategy.

Keywords: non-sequential structure alignment methods; protein structures; measurement.

1. Introduction

Protein structure alignment approaches are of great importance for the analysis of protein function, structure and evolution. A large number of structure alignment programs have been developed in the recent years to find a solution to this NP-hard problem (see for instance [1]). A representative set of sequential alignment programs is DaliLite [2], K2 [3], CE [4] and TM-align [5]. In addition to these sequential structure alignment methods, also non-sequential structure alignment methods have been developed i.e. TOPOFIT [6], MASS [7], GANGSTA [8], GANGSTA+ [9] (see Table 1) and others [10, 11] to mention just a few. Due to the large number of available programs and the variety of underlying concepts and strategies, selecting the "best" method for a certain application is difficult. Generally, a comparison of protein structure alignment algorithms is carried out on the basis of reasonably difficult and representative protein pair benchmark sets [12]. The resulting protein structure alignments obtained for a selected set of protein pairs can be evaluated with a wide variety of protein structure alignment scoring functions i.e. Z-Score [2], TM-score [5] and SAS [13]. These scoring functions have in common to consider two major properties of a protein structure alignment, which are the number of aligned residue pairs between the protein pair and the corresponding root mean square deviation (RMSD) of the C_α atoms of the two structures. Many scoring functions consider also other characteristic quantities such as total number of unaligned sequence segments (gaps) or the amino acid identities of aligned residues to achieve an improved and more reliable measure of the alignment quality. Similar to sequence alignment approaches [14, 15] a p-value (see also Z-Score [2]) can be evaluated, defining the probability that an unrelated protein structure pair obtains by chance a specific score. Scoring functions can be constructed to be insensitive versus the size of proteins, as for instance the TM-score [5]. The biological relevance of a scoring function can be analyzed by probing its ability to recognize protein families or alternative relations for a given set of proteins [16]. For these tests high-quality protein structure databases like e.g. CATH [17, 18] and SCOP [19] are used as a reference. Kolodny et al. [13] studied the problem to find scoring functions that are suitable to compare and measure the quality of protein structure alignments obtained with different algorithms and proposed a set of four scoring functions i.e. SI, MI, SAS and GSAS. To judge the performance of different alignment tools the benchmark of protein structure pairs and the choice of the scoring function are critical.

An issue, which is particularly important for the evaluation of non-sequential structure alignment methods, is the underlying strategy of residue assignment. Although often disregarded, this issue can have a significant influence on the results of the alignment algorithms as we will show in the present study. To illustrate these influences, firstly we selected a minimal set of five individual residue assignment options. They comprise the following features: (α) to align residues located everywhere in the structure or only within secondary structure elements (SSE) (i.e. α-helix and β-strand); (β) to align two SSEs also in reverse orientation (i.e. N-terminus of the SSE from one protein is aligned on the C-terminus of the corresponding SSE of the second protein); (γ) to make only unique or also shared (fuzzy) residue assignments. In the latter case the residue of one protein can be assigned to more than one residue in the other protein; (δ) to assign only residues belonging to the same SSE type or to ignore the SSE type in these assignments; (ϵ) to assign residues without gaps or to assign also isolated residue pairs. These residue assignment options [(α)-(ϵ)] are combined to a set of five residue assignment strategies [(1) – (5)] as defined in Table 2.

To analyze the influences of these different alignment strategies [(1) – (5)], we implemented them in GANGSTA+ [9]. We like to point out that this study can also be done with several other protein structure alignment tools mentioned above. We use GANGSTA+ in this application since with our own method we have the procedures better under control. As a benchmark set five circular permuted protein structure pairs were considered. To obtain reference values for the residue assignment strategies [(2) – (5)], we applied the non-sequential protein structure alignment approaches TOPOFIT [6], MASS[7], GANGSTA+ [9] and GANGSTA [8] (see Table 1) on the same five protein pairs. We like to emphasize that this work does not intend to compare the overall performances of these non-sequential structure alignment methods, but to illustrate the effect of their specific residue assignment strategies on the number of aligned residues and the RMSD.

Firstly, the results reveal an essential effect of the implemented residue assignment strategies on the number of aligned residues and the RMSD. Secondly, these results align well to the reference values, observed by applying the corresponding protein structure alignment method for each residue assignment strategy. Concluding, this shows that any performance comparison of protein structure alignment methods has to consider the variations in the underlying residue assignment strategies. Furthermore, considering these variations can explain observed performance differences between entirely different non-sequential structure alignment algorithms.

2. Methods

2.1. *Benchmark of circular permuted protein structure pairs*

The benchmark dataset for the comparison of non-sequential protein structure alignment algorithms consists of five circular permuted protein pairs. These are the protein pairs with PDB [20] id (a) 1RIN [21] / 2CNA [22], (b) 1GLH [23] / 1CPN [24], (c) 1EXG [25] / 1TUL [26], (d) 1RHG [27] / 1BCF [28], (e) 1IHW [29] / 1SSO [30] (see Table 1). The same dataset is used to illustrate the effect of different residue assignment strategies with GANGSTA+ (see Table 2 and Figure 1).

2.2. *Non-sequential protein structure alignment tools*

Four non-sequential protein structure alignment approaches were considered in this study (listed in Table 1). GANGSTA [8] operates hierarchically on two stages. First, a genetic algorithm is employed to optimize the SSE assignment between two protein structures by maximizing a contact map overlap based on the specific GANGSTA objective function (GOF). Second, the SSE assignment is transferred to the residue level. GANGSTA ignores segments with loop and coil structure and operates in sequence direction only, i.e. SSEs in a protein pair are not aligned in reverse sequential order. GANGSTA assigns residues only, if they belong to the same SSE type and allows no gaps in the same SSE. GANGSTA+ [9] is the successor of GANGSTA and employs an efficient combinatorial approach for the SSE assignment to maximize the GOF. Thus, both methods share the same optimization target on the SSE level. GANGSTA+ transfers the SSE assignment to the residue level by a heuristic point matching potential function, which minimizes the distances between aligned residue pairs. In addition, GANGSTA+ refines and extends the residue assignment from SSEs to loop and coil segments and can also detect similarities between SSEs assigned in reverse sequential order. With default setting GANGSTA and GANGSTA+ assign sequential patches of residue pairs belonging to the same SSE types only. Both allow database searches. TOPOFIT [6] is based on geometric hashing. Its online service allows database searches. TOPOFIT aligns residues including loops and coils in both sequence directions. Single residue alignment is allowed. It does not distinguish between SSE types, when maximizing the geometrical similarities of a protein structure pair. MASS [7] is an algorithm designed for multiple protein structure alignment. Residue assignments occur similar to TOPOFIT, i.e. both include loops and coils and are able to detect structure similarities in opposite sequence orientation. In addition, both algorithms allow gaps in the residues aligned to the same SSE pair. In contrast to TOPOFIT, MASS distinguishes the SSE types of the aligned residues, ensuring that they are aligned type consistently.

3.   Results

3.1. *Application of different non-sequential protein structure alignment tools*

We applied TOPOFIT [6], MASS [7], GANGSTA [8] and GANGSTA+ [9] on the benchmark of five circular permuted protein pairs. TOPOFIT and MASS yielded an average SAS of 2.0 Å [SAS definition [13]: SAS = (100*RMSD) / $N_{aligned}$ with $N_{aligned}$ number of aligned residues and RMSD in Å] (see [6, 7] for details on computing time). With default setting GANGSTA+ yielded a slightly larger average SAS = 2.5 Å. The CPU time required by GANGSTA+ is less than 1s [1.6 GHz AMD/OPTERON] per protein structure pair. For GANGSTA the structure alignment results took about 10s per protein pair on average [1.6 GHz AMD/OPTERON]. With its intrinsic GOF and scoring function for the residue assignment GANGSTA employed a different protein structure alignment strategy designed to detect more distant structural similarities and is not extending the residue assignment into the loop segments. In the present benchmark the degree of structure similarity was high, but GANGSTA performed less well yielding an average SAS of 5.5 Å only. Table 1 depicts the resulting RMSD and $N_{aligned}$ of the considered methods for the benchmark of five protein structure pairs.

Table 1. Protein structure alignment results on a benchmark set of five protein structure pairs.

	protein 1	protein 2	TOPOFIT	MASS	GANGSTA+	GANGSTA
a)	1RIN:A(180)[a]	2CNA:_(237)[a]	152/1.09 [b]	164/1.2[b]	147/1.36 [b]	56/0.78 [b]
b)	1GLH:_(214)	1CPN:_(208)	206/0.49	206/0.49	205/0.48	97/0.26
c)	1EXG:_(110)	1TUL:_(102)	52/1.79	60/1.9	43/2.02	41/3.5
d)	1RHG:A(145)	1BCF:A(158)	109/1.40	106/1.7	108/1.38	100/2.5
e)	1IHW:A (52)	1SSO:_(62)	35/1.47	39/1.7	31/1.70	16/2.4

[a] PDB id: domain id according to SCOP (number of residues)
[b] number of aligned residues / RMSD in Å

3.2. *Application of different residue assignment strategies with GANGSTA+*

We generated protein structure alignments for the benchmark of the five circular permuted protein structure pairs with GANGSTA+ using five different residue assignment strategies, listed in Table 2. Hereby, we intended to capture the major differences in the residue assignment strategies of the non-sequential structure alignment methods employed in the section above. The resulting number of aligned residues and the RMSD values for the five protein pairs are displayed in Figure 1. Strategy (1) comprises the least constrained residue assignment, thus yielding the smallest (optimal) average SAS = 1.7 Å. Relatively small SAS = 2.0 Å were also obtained with strategy (2) and (3) where residue pairs were assigned uniquely as done by TOPOFIT and MASS. Additional constraints requiring that residues belonging to the same pair of SSEs must be aligned without gaps correspond to the default setting of GANGSTA+ [9]. The SAS values (average SAS = 2.5 Å) for this strategy (4) are shown as orange colored symbols in Figure 1. The alignment can be further restricted such that coils are ignored and SSEs cannot be aligned in reverse orientation [strategy (5)], which corresponds to the default setting of GANGSTA and yields the average SAS = 4.0 Å. Strategies (2) and (3) differ only in the type consistency of the aligned SSEs, which is fulfilled for (3) but not for (2). In the present application this algorithmic difference did not effect the results, while in other studies significant effects were observed varying this option [9]. Only, the results for the protein structure pair (d) 1RHG / 1BCF remained invariant for different residue assignment strategies. This contrasts with the results for the other four protein structure pairs, which exhibit large variations in the number of aligned residues and the corresponding RMSD (see Figure 1).

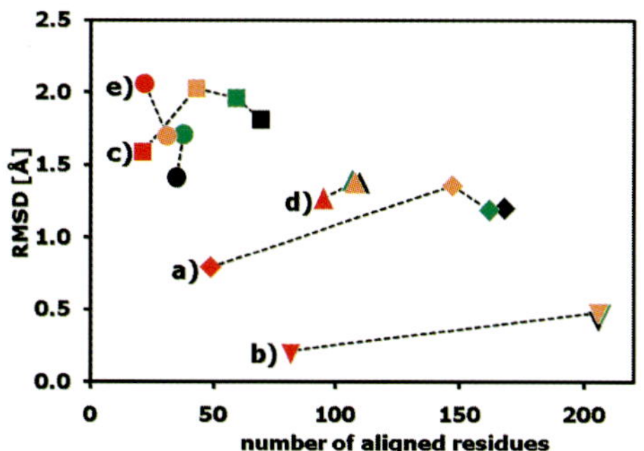

Fig. 1. Results of protein structure alignments with GANGSTA+ using different residue assignment strategies see Table 2 [black (1), green (2,3), orange (4), red (5)] on the benchmark set of five protein structure pairs listed in Table 1 (dashed lines to guide the eye). The protein pairs are (a) 1RIN [21] / 2CNA [22] (diamonds), (b) 1GLH [23] / 1CPN [24] (inverted triangles), (c) 1EXG [25] / 1TUL [26] (squares), (d) 1RHG [27] / 1BCF [28] (triangles), (e) 1IHW [29] / 1SSO [30] (circles).

Table 2. Different residue assignment strategies used for non-sequential protein structure alignment. A detailed description of the assignment options (α)-(ε) is given in the introduction.

residue assignment options /strategies	1	2	3	4	5
(α) includes loops and coils in the alignment	x	x	x	x	
(β) detects sequence reverse alignments	x	x	x	x	
(γ) assigns residue pairs uniquely		x	x	x	x
(δ) assigns residues of same SSEs type			x	x	x
(ε) assigns residues gapless on SSE				x	x

4.   Conclusion

This study illustrates that a balanced evaluation of protein structure alignments generated with commonly used methods can be difficult due to algorithmic differences regarding the assignment strategy of residues. Bringing these differences in line is essential for a fair comparison of protein structure alignment tools. We applied the originally developed alignment program GANGSTA [8] on a set of five circular permuted protein structure pairs. GANGSTA needs about 10s CPU time [AMD/OPTERON at 1.6 GHz] per protein pair, but yielded the largest average SAS value compared to the three other considered non-sequential structure alignment algorithms. There are mainly two constraints used by GANGSTA, which contribute to the larger SAS values. GANGSTA assigns only residues belonging to SSEs and strictly ignores residues in loops and coils. Another reason for the reduced performance of the genetic algorithm in the original GANGSTA method is its tendency to not fully explore the search space. Therefore, the genetic algorithm in GANGSTA was replaced by a more reliable combinatorial approach in GANGSTA+ [9].

The residue assignment strategies used by the other alignment tools are generally less constrained than in GANGSTA [8]. The least constrained strategy (1) of fuzzy residue assignment (a single residue of one protein is assigned to more than one residue in the other protein) should be analyzed in more detail in future work. This strategy might improve detection of conserved residues in protein structure alignment. However, a comparison of the alignment quality of protein structures obtained from different alignment tools, based on the number of aligned residues and RMSD can be particularly misleading in this case. GANGSTA+, TOPOFIT and MASS performed similar on the considered benchmark (average SAS between 2.0 Å and 2.5 Å), yielding consistently very good protein structure alignments, although each of the alignment tools is based on entirely different optimization algorithms. These results illustrate the efficiency of currently available protein structure alignment approaches in solving non-sequential structure alignment problems. In a second case study, we generated protein structure alignments of the same benchmark set using GANGSTA+ with varying residue assignment strategies. These strategies have a significant effect on the number of aligned residues and the corresponding RMSD (see Figure 1). One has to keep in mind that with its default settings GANGSTA+ aims to generate alignments with consistent and complete SSE assignments and ignores loops and coils in the initial stage of optimizing protein structure alignments (see details in [9]). However, we implemented different residue assignment strategies as employed by other non-sequential structure alignment methods (see Table 2). In this way GANGSTA+ seems to reproduce the results of TOPOFIT and MASS for the considered benchmark faithfully. Allowing shared residue assignments, GANGSTA+ yields a further decrease of the resulting SAS values yielding an average SAS of 1.7 Å.

This study underlines the strength of current non-sequential protein structure alignment tools. These are capable to detect sophisticated similarities with SSEs in reverse orientation, circular permuted protein pairs and can consider shared residue assignments. However, these capabilities cause difficulties, when carrying out performance comparisons between different methods. Comparing the number of aligned residues or the RMSD can be very misleading, even if the same protein pair is considered. Since, these two properties are very essential to almost every scoring function a comparison of different methods can be carried only if the underlying residue assignment strategies are taken into account.

Acknowledgments

This work was supported by the International Research Training Group (IRTG) on "Genomics and Systems Biology of Molecular Networks" (GRK1360, Deutsche Forschungsgemeinschaft (DFG)).

References

[1]   Wikipedia, [http://en.wikipedia.org/wiki/Structural_alignment_software].
[2]   Holm, L., Park, J., DaliLite workbench for protein structure comparison, *Bioinformatics*, 6:566-7, 2000.
[3]   Szustakowski, J., Weng, Z., Protein structure alignment using a genetic algorithm, *Proteins*, 38(4):428-440, 2000.
[4]   Shindyalov, I. N., Bourne, P. E., Protein structure alignment by incremental combinatorial extension (CE) of the optimal path, *PEDS*, 11:739-747, 1998.
[5]   Zhang, Y., Skolnick, J., TM-align: A protein structure alignment algorithm based on TM-score, *Nucleic Acids Research*, 33:2302-2309, 2005.
[6]   Ilyin, V., Abyzov, A., Leslin, C., Structural alignment of proteins by a novel TOPOFIT method, as a superimposition of common volumes at a topomax point, *Protein Science*, 13:1865-1874, 2004.
[7]   Dror, O., Benyamini, H., Nussinov, R., Wolfson, H. J., MASS: multiple structural alignment by secondary structures, *Bioinformatics*, 19:95-104, 2003.
[8]   Kolbeck, B., May, P., Schmidt-Goenner, T., Steinke, T., Knapp, E. W., Connectivity independent protein-structure alignment, *BMC Bioinformatics*, 7(510), 2006.
[9]   Guerler, A., Knapp, E. W., Novel protein folds and their non-sequential structural analogs, *Protein Science*, 17:1374-1382, 2008.
[10] Bystroff, Y., Non-sequential structure-based alignments reveal topology-independent core packing arrangements in proteins, *Bioinformatics*, 7:1010–1019, 2005.
[11] Teichert, F., Bastolla, U., Porto, M., SABERTOOTH: protein structural alignment based on a vectorial structure representation, *BMC Bioinformatics*, 8(425), 2007.
[12] Fischer, D., Elofsson, A., Rice, D., Eisenberg, D., Assessing the performance of fold recognition methods by means of a comprehensive benchmark, *Pac. Symp. Biocomput.*, 1:300-318, 1996.

[13] Kolodny, R., Koehl, P., Levitt, M., Comprehensive Evaluation of Protein Structure Alignment Methods, *Journal of Molecular Biology*, 346:1173-1188, 2005.

[14] Altschul, S. F., Madden, T. L., Schaffer, A. A., Zhang, J., Zhang, Z., Miller, W., Lipman, D. J., Gapped BLAST and PSI-BLAST: a new generation of protein database search programs, *Nucleic Acids Research*, 25(17):3389-3402, 1997.

[15] Henikoff, S., Henikoff, J. G., Amino acid substitution matrices from protein blocks, *PNAS*, 89:10915-10919, 1992.

[16] Day, R., Beck, D. A. C., Armen, R. S., Daggett, V., A consensus view of fold space: Combining SCOP, CATH, and the Dali Domain Dictionary, *Protein Science*, 12:2150-2160, 2003.

[17] Orengo, C. A., Michie, A. D., Jones, S., Jones, D. T., Swindells, M. B., Thornton, J. M., CATH- A Hierarchic Classification of Protein Domain Structures, *Structure*, 5(8):1093-1108, 1997.

[18] Pearl, F. M. G., Lee, D., Bray, J. E., Sillitoe, I., Todd, A. E., Harrison, A. P., Thornton, J. M., Orengo, C. A., Assigning genomic sequences to CATH, *Nucleic Acids Research*, 28(1):277-282, 2000.

[19] Murzin, A. G., Brenner, S. E., Hubbard, T., Chothia, C., SCOP, *J. Mol. Biol.*, 247:536-540, 1995.

[20] Berman, H. M., Westbrook, J., Feng, Z., Gilliland, G., Bhat, T. N., Weissig, H., Shindyalov, I. N., Bourne, P. E., The Protein Data Bank, *Nucleic Acids Research*, 28:235-242, 2000.

[21] Rini, J. M., Hardman, K. D., Einspahr, H., Suddath, F. L., Carver, J. P., X-ray crystal structure of a pea lectin-trimannoside complex at 2.6 A resolution, *J.Biol.Chem.*, 268:10126-10132, 1993.

[22] Reeke, G. N., Becker, J. W., Edelman, G. M., The covalent and three-dimensional structure of concanavalin A. IV. Atomic coordinates, hydrogen bonding, and quaternary structure, *J.Biol.Chem.*, 250:1525-1547, 1975.

[23] Keitel, T., Meldgaard, M., Heinemann, U., Cation binding to a Bacillus (1,3-1,4)-beta-glucanase. Geometry, affinity and effect on protein stability, *Eur. J. Biochem.*, 222:203-214, 1994.

[24] Hahn, M., Piotukh, K., Borriss, R., Heinemann, U., Native-like in vivo folding of a circularly permuted jellyroll protein shown by crystal structure analysis, *Proc.Natl.Acad.Sci.USA*, 91:10417-10421, 1994.

[25] Xu, G. Y., Ong, E., Gilkes, N. R., Kilburn, D. G., Muhandiram, D. R., Harris-Brandts, M., Carver, J. P., Kay, L. E., Harvey, T. S., Solution structure of a cellulose-binding domain from Cellulomonas fimi by nuclear magnetic resonance spectroscopy, *Biochemistry*, 34:6993-7009, 1995.

[26] Holden, H. M., Wesenberg, G., Raynes, D. A., Hartshorne, D. J., Guerriero, V., Rayment, I., Molecular structure of a proteolytic fragment of TLP20, *Acta Crystallogr.*, 52:1153-1160, 1996.

[27] Hill, C. P., Osslund, T. D., Eisenberg, D., The structure of granulocyte-colony-stimulating factor and its relationship to other growth factors, *Proc. Natl. Acad. Sci. USA*, 90:5167-5171, 1993.

[28] Frolow, F., Kalb, A. J., Yariv, J., Structure of a unique twofold symmetric haem-binding site, *Nat.Struct.Biol.*, 1:453-460, 1994.

[29] Lodi, P. J., Ernst, J. A., Kuszewski, J., Hickman, A. B., Engelman, A., Craigie, R., Clore, G. M., Gronenborn, A. M., Solution structure of the DNA binding domain of HIV-1 integrase, *Biochemistry* 34:9826-9833, 1995.

[30] Baumann, H., Knapp, S., Lundback, T., Ladenstein, R., Hard, T., Solution structure and DNA-binding properties of a thermostable protein from the archaeon Sulfolobus solfataricus, *Nat.Struct.Biol.*, 1:808-819, 1994.

ACTIVE PATHWAY IDENTIFICATION AND CLASSIFICATION WITH PROBABILISTIC ENSEMBLES

TIMOTHY HANCOCK

timhancock@kuicr.kyoto-u.ac.jp

HIROSHI MAMITSUKA

mami@kuicr.kyoto-u.ac.jp

Bioinformatics Center, Institute for Chemical Research, Kyoto University, Kyoto, Japan

A popular means of modeling metabolic networks is through identifying frequently observed pathways. However the definition of what constitutes an observation of a pathway and how to evaluate the importance of identified pathways remains unclear. In this paper we investigate different methods for defining an observed pathway and evaluate their performance with pathway classification models. We use three methods for defining an observed pathway; a path in gene over-expression, a path in probable gene over-expression and a path of most accurate classification. The performance of each definition is evaluated with three classification models; a probabilistic pathway classifier - HME3M, logistic regression and SVM. The results show that defining pathways using the probability of gene over-expression creates stable and accurate classifiers. Conversely we also show defining pathways of most accurate classification finds a severely biased pathways that are unrepresentative of underlying microarray data structure.

Keywords: Classification; Markov model; Mixture of Experts; Metabolic Pathway.

1. Introduction

A metabolic pathway is a small section of the overall metabolic network that comprises a connected series of chemical reactions which are known to perform a specific function (Figure 1). On the left hand side of Figure 1 we present a simplified diagram of a metabolic pathway of similar structure to those found within on line metabolic network databases such as KEGG [4]. The metabolic compounds of the network are $[C_1, \ldots, C_4]$ are connected by reactions $[R00001, R00002, R00003, R00004]$ which in turn are catalyzed by a set of genes $[G_1, \ldots G_8]$. To define a pathway through the network we have flagged C_1 as the start compound and C_4 as the end. Observations of the genes within Figure 1 come most commonly from microarray experiments which measure the expression levels of the genes within the network. Therefore it is convenient to transform the reaction-compound network shown on the left hand side of Figure 1 into the gene-compound network shown on the right. A pathway through the gene-compound network from C_1 to C_4 can now be defined as a sequence of genes denoting the edges between compounds.

Determining what level of gene expression indicates an active network edge is not

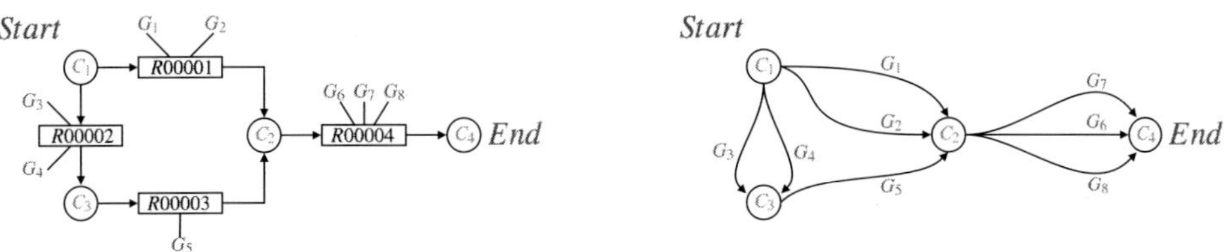

Fig. 1: Diagrammatic representation of a metabolic network.

simple due to the high level of noise inherent within microarrays. By far the most common approach for specifying active genes rely on summary statistics such as fold change and z-scores. However using simple summary statistics to denote an active edge reduces number of observations on the activity of the network down to a single value for each response label. Reducing all microarray observations to a single value could potentially remove important subtle changes in expression structure that can affect which path is taken. Another common approach is to employ a supervised technique to pick genes from a metabolic network that have expression structure useful for classifying a response. However the process of selecting important genes is likely to break network relationships between the genes leading to a disconnected solution which makes any biological pathway interpretation difficult.

In this paper we consider the problem of what is an appropriate definition for an active edge. To ensure that the full structure of the underlying microarray is represented we require a solution that can extract observed pathways from each microarray experiment. We consider three definitions of an active edge:

(1) A high value of scaled gene expression data.
(2) A high probability found within an empirical cumulative distribution function computed for all genes within the network.
(3) A low entropy of the classified label of microarray for each experiment computed by individual gene classifiers.

We perform a comparative analysis across the three gene activity criteria using standard classification models; support vector machines and logistic regression as well as an explicit classifier of metabolic pathways, HME3M [2]. The success of each gene activity criteria is evaluated both in terms of any observed bias in the number pathways extracted and in terms of classification accuracy and stability. In the next section we describe the pathway extraction procedure and each edge activity criteria. We then describe our pathway classifiers and compare the performance of each method on real biological data.

2. Pathway Extraction

In our experiments we extract all valid pathways from each microarray experiment that are observed from the start to the end compounds of a specified network. To do this we treat each microarray experiment, x_i as a single observation of the activity of all genes within a network. For each x_i we also have a response label y_i denoting the experimental conditions. Then using a pre-specified criteria to determine which genes are active within x_i we extract all possible paths from the start node to the end node and label each path with y_i. For example in the example network in Figure 1 if we determine that for a given experiment, x_i, the genes $[G_1, G_4, G_5, G_6]$ are active then 2 possible paths can be extracted and represented as binary strings $[G_1, G_6] = [1, 0, 0, 0, 0, 1, 0, 0]$ and $[G_4, G_5, G_6] = [0, 0, 0, 1, 1, 1, 0, 0]$. Both of these paths are then given the response label y_i. If all eight genes within Figure 1 are found to be active then a total of 12 possible pathways can be extracted. From each microarray experiment we extract all possible paths and their response labels and augment them into our final pathway dataset.

Extracting all possible paths observed over all microarray experiments presents two major issues for an further analysis. Firstly how to determine the criteria which denotes active genes within each microarray experiment, and secondly how to handle the inevitable duplicate pathways that will be extracted. In our experiments we focus on the first issue of assessing gene activity, but also address what effect duplicate paths by first including them within the analysis and then monitoring the performance change observed after their removal. We now describe the three gene activity criteria under consideration.

3. Gene Activity Criteria

3.1. *Scaled Expression*

The z-scaling or normalization of gene expression data is a standard preprocessing step performed before most microarray analysis and is simply,

$$z_i = \frac{x_{ij} - \bar{x}_j}{s_{x_j}} \tag{1}$$

where $\bar{x}_j$ is the mean of all expressions for gene x_j and s_{x_j} is the standard deviation. The effect of normalization is that each gene will have a mean of $\bar{x}_j = 0$ and a standard deviation of $s_{x_j} = 1$. Normalization therefore brings all gene expressions down to a relative scale allowing for easy comparison. However normalization using the z transformation does assume that the distribution of each gene is independent and normally distributed.

3.2. *Empirical CDF*

To overcome the assumptions of a z-score non-parametric procedures such as ranking the gene expressions are often employed. In this paper we rank gene expressions based on their position within the empirical Cumulative Distribution Function (CDF) of all genes. Although an empirical CDF is equivalent to a standard ranking procedure there are two major advantages for its application; firstly that the empirical CDF provides a function capable of looking up new gene expression observations without altering the ranking, and secondly by using an empirical CDF the gene expressions are transformed to probabilities where a probability of 0.5 provides an intuitive of gene activity definition.

To compute the empirical CDF over all genes we first ignore all gene structure and extract a simple vector of individual expressions. Then we compute the probability of each individual expression value by,

$$P(x_{ij}) = \int_{-\infty}^{x_{ij}} P(t).dt = \frac{\text{Number}[x < x_i ij]}{\text{Number}[x]} \tag{2}$$

where $\text{Number}[x < x_i]$ indicates the number of individual expressions below x_{ij}. Clearly the final empirical CDF probability of each observation unlike the z-score does not assume any specific distribution.

3.3. *Entropy*

Both the z-score and empirical CDF criteria are unsupervised naive transformations on the gene expression. We also include a model dependent transformation to assess the ability of using a supervised technique to select active genes within a pathway. To do this we perform a logistic regression on each variable individually and extract the posterior probabilities of classification. From the posterior probabilities we compute the entropy of the classification for each observation,

$$\text{Entropy}(x_{ij}) = \sum_{y_l \in y} p(x_{ij}|y = y_l) \log p(x_{ij}|y = y_l) \tag{3}$$

where y_l is a label within the response y and $p(x_{ij}|y = y_l)$ are the posterior probabilities for each observation x_{ij} and each response label y_l. The entropy of the posterior probabilities is an indication of how certain each logistic regression is on its assigned label. The entropy of a random assignment where $p(x_{ij}|y = y_l) = 0.5 \; \forall \; y_l$ is approximately -0.6931 and as the certainty of the classification increases the entropy tends to 0. Therefore if all labels are classified well the entropy should be small, and a small entropy is an indication that the network structure is well defined for a particular observation. Entropy however does assume a particular model for each gene, and clearly the results will be dependent on the choice of model. In this paper we select logistic regression because of the smooth transition between classes that is provided by the assumed logistic curve.

3.4. *Comparison of Activity Criteria*

In Figure 2 we present a example of the effect of each of our gene activity criteria on 5 simulated genes connected in a single path from the gene 1 to gene 5 (left to

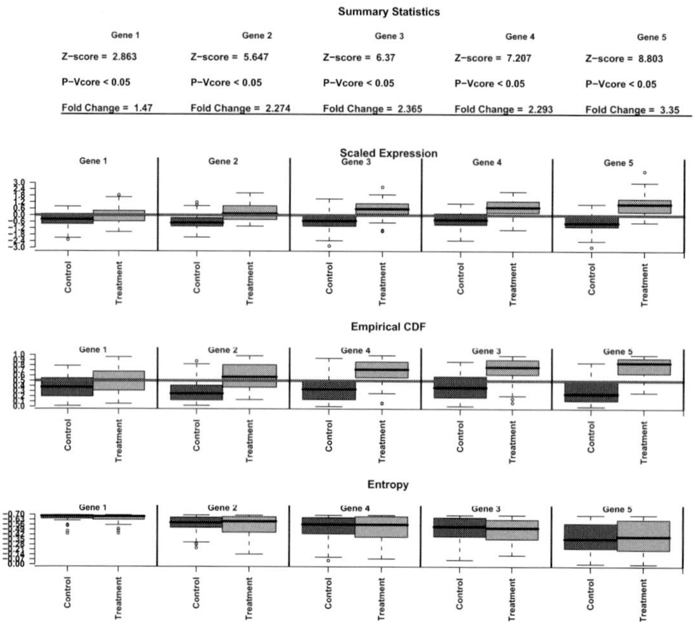

Fig. 2: Boxplots displaying the effect of different gene activity criteria.

right). From left to right we have simulated a subtle increase in gene expression observed in the treatment group. This is observed in the z-scores and corresponding P-values as well as expression fold changes computed for each gene presented at the top of Figure 2. It is clearly seen the z-scores for all genes find the treatment group to be significantly over-expressed at a P-value < 0.05 and therefore we observe the overall path to be active.

However for individual gene activity criteria marked differences in the structure of each transformed gene are observed. For the scaled expression values we observe that the range of expressions within each response label is rather small. The red line across all box plots is drawn at a z-score of 0 where all observations above this line intuitively correspond to active gene expression. However because small expression range we observe that a subtle change in this tolerance value will create a large difference in the number of observations deemed to be active. Conversely for the empirical CDF criteria we observe that the range of expression within each response label is broader but the structure of the overall expression profile is maintained. This implies that if we slightly move the gene activity tolerance a similar gene expression profile will be obtained and therefore we expect pathway extraction using empirical CDF to be more stable than the standard scaled expression approach.

The entropy criteria however presents the opposite interpretation, where an active gene is determined by a low entropy score. It is clear from Figure 2 that there is a large difference in the entropy profiles from the genes showing accurate performances compared to G_1 where inaccurate classifications are observed. The inaccurate classifications of G_1 are in contrast to the summary statistics which show the entire path to be active. This implies that a single inaccurate gene classifier may break the pathway even though the data indicates the pathway to be active. To overcome this problem the entropy tolerance may need to be decreased to unrealistic levels which will result in the number paths extracted to become large and cause serious interpretation problems and performance penalties.

4. Pathway Classification

To compare the performance of each gene activity criteria we analyze the pathways with benchmark classification models penalized logistic regression (PLR) [6] and Support Vector Machines (SVM) with linear, polynomial (degree $= 3$) and radial basis kernels [1]. Furthermore to analyze the specific pathway differences found by each criteria we also compare the model performances with a specialized pathway classification model HME3M [2].

4.1. *Hierarchical Mixture of Markov Experts (HME3M)*

HME3M is a Hierarchical Mixture of Experts (HME) [3] which uses a Markov mixture model 3M [5] to first cluster the pathways. Then supervision of this cluster analysis is performed by training experts (classification models) on the pathways found at each cluster. This process however is not performed as two discrete steps, but is iteratively optimized with an EM algorithm. The EM optimization passes information from the experts back into the pathway clustering algorithm and vice versa. This flow of information from the experts into the clustering algorithm allows for clusters to be found that will provide optimal classification performances at each expert. Therefore HME3M is an probabilistic ensemble of experts where each expert evaluates the classification accuracy of each pathway cluster.

Combing a HME with the 3M model produces the HME3M likelihood,

$$p(y|x) = \sum_{m=1}^{M} \pi_m p(y|x, \beta_m) p(c_1|\theta_{1m}) \prod_{t=2}^{T} p(c_t, x_t|c_{t-1}; \theta_{tm}) \,. \tag{4}$$

The parameters of (4) are estimated with the EM algorithm by defining the responsibilities variable h_{im} to be the probability that a sequence i belongs to component m, given x, θ_m, β_m and y and using the following E and M steps:

E-Step: Define the responsibilities h_{im}:

$$h_{im} = \frac{\pi_m p(m|x_i, \theta_m) p(y_i|x_i, \beta_m)}{\sum_{m=1}^{M} \pi_m p(m|x_i, \theta_m) p(y_i|x_i, \beta_m)} \tag{5}$$

M-Step: Estimate the Markov mixture and expert model parameters:

(1) Estimate the mixture parameters:

$$\pi_m = \frac{\sum_{i=1}^{N} h_{im}}{\sum_{m=1}^{M} \sum_{i=1}^{N} h_{im}} \quad \text{and} \quad \theta_{tm} = \frac{\sum_{i=1}^{N} \delta(x_{it} = 1) h_{im}}{\sum_{i=1}^{N} h_{im}} \tag{6}$$

(2) Estimate the expert parameters:

We model our experts with a Penalized Logistic Regression (PLR) [6],

$$l(\beta_m|h_{im}) = \arg\max_{\beta_m} \left\{ \sum_{i=1}^{N} h_{im}\left(y_i\beta_m^T x_i + log(1 + e^{\beta_m^T x_i})\right) - \frac{\lambda}{2}|\beta_m|^2 \right\} \qquad (7)$$

where X are the paths, y is a binary response variable, h_{im} are HME3M responsibilities for path i in component m, and β_{tm} are the PLR coefficients.

5. Experiments

We perform our experiments on a subsection of the glycolysis pathway for *Arabidopsis thaliana* extracted from KEGG [4] using gene expression information from the AltGenExpress database [7]. The pathway shown in Figure 3 starts at Alpha-D-Glucose and finishes at Pyruvate, contains 41 edges, and for a single observation there are 103680 possible paths. We extract observations from the microarray for two classes; *"rosette leaf"* ($n = 21$) and *"flower"* ($n = 15$) and specify "flower" to be target class ($y = 1$) and "rosette leaf" to be the comparison class ($y = 0$).

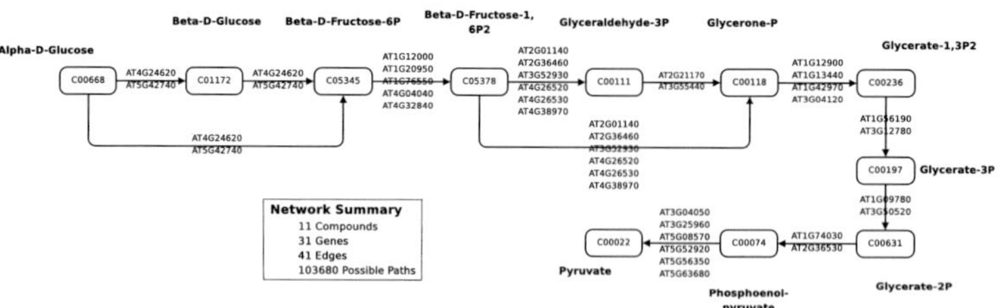

Fig. 3: *Arabidopsis thaliana* glycolysis pathway from Alpha-D-Glucose to Pyruvate.

We select this pathway as in our previous research [2] we used HME3M to show that there is a clear differential pathway expression that can be used to create a stable classifier of flower gylcolysis pathways. However in [2] we only employed a scaled expression tolerance, and did not consider different pathway definitions or duplicate pathways. In this research we focus on the effect of the pathway extraction procedure to see if the same pathways found in [2] can be more efficiently found using different pathway definitions and further assess the effect of removing duplicate pathways from the dataset.

6. Results: Pathway Extraction

We vary the tolerance for each gene activity criteria over a range such that the number of paths extracted from each class for each criteria would be as similar as possible. We vary the scaled expression tolerance between $[-0.14, 0.14]$, the

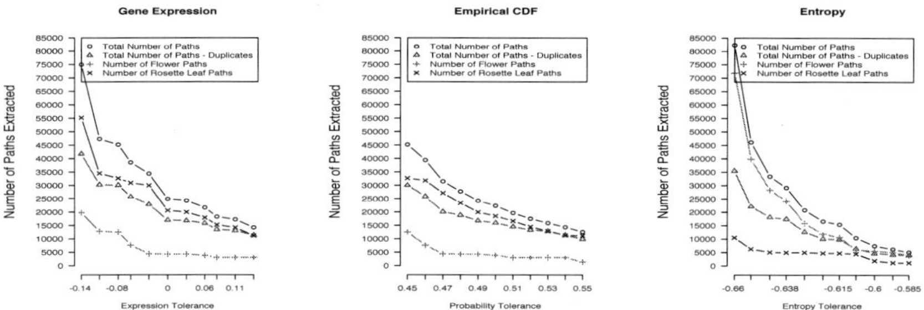

Fig. 4: Number of pathways extracted using different gene activity criteria.

empirical CDF tolerance between $[0.45, 0.55]$ and the entropy tolerance between $[0.54, 0.6]$. The results are presented in Figure 4. In Figure 4 the black line represents the total number of paths extracted and the red line represents the number of paths after duplicate pathways have been removed. The blue line represents the number of paths extracted from the rosette leaf experiments and the green line represents the number of number of paths extracted from the flower experiments.

The pathway extraction profiles agree with the results found in the simulation experiments (Figure 2). We see that the scaled expression criteria at low tolerance values extracts a large number of paths and then small increases in this tolerance rapidly decreases the number of pathways found. In contrast the empirical CDF begins at a smaller number of extracted paths and a constant rate of decrease in the number of pathways extracted is observed as the tolerance level increases.

For the entropy criteria it is clear that the extracted number of paths will dramatically decrease with a small increase in tolerance. The large decrease in pathways observed requires the setting of a small and imprecise entropy tolerance range, $[-0.66, -0.58]$, such that observations from both response labels could be extracted. We also observe that decreasing the entropy is clearly biased to the flower experiments. This bias towards the flower experiments is because they are predicted less accurately by some individual gene logistic regressions. These inaccurate predictions of the flower experiments results means they have a entropy and therefore decreasing the entropy tolerance results in large number of flower pathways to be extracted. This clear bias in the number of extracted pathways towards the flower experiments is likely seriously effect performance and stability of each classifier.

7. Results: Pathway Classification

For the HME3M model over all experiments we set the number of components to be four ($M = 4$) and $\lambda = 1$ to agree with the parameters used in [2]. For all models to cope with the number of pathways extracted inverse 20-fold cross validation is employed and each model is compared based on its test set correct classification rate (CCR). The results are presented in Figure 5.

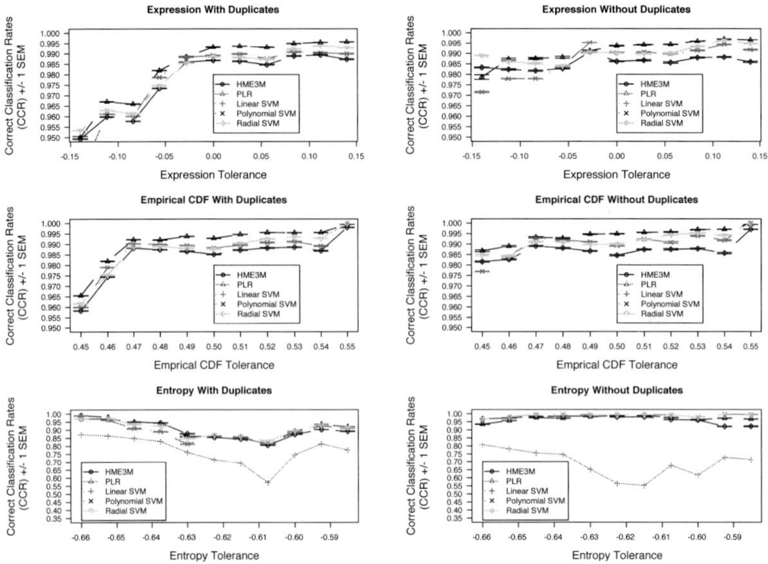

Fig. 5: Performance of each classification model on the pathways extracted with different gene activity criteria and with duplicate pathways removed.

From Figure 5 for the scaled expression models with duplicate pathways the optimal performance is reached after a tolerance of 0 is exceeded. In comparison the empirical CDF models with duplicates pathways reach the same optimal performance earlier in the experiments, after a tolerance of 0.47 is exceeded. For both criteria the once the optimum is reached all models maintain their performance with further increases in tolerance. These observations clearly show that the empirical CDF is extracting a more stable a set of pathways with a structure that is resistant to reasonable changes in tolerance. In contrast the performance profile for the entropy criteria with duplicates pathways is not stable. This is most likely due to the severe bias in the number of pathways extracted for the flower experiments. This severe bias towards the flower observation is artificially inflating the prediction results and removing this bias can be used to explain the decrease in model performances decrease as the entropy tolerance is increased.

Over all models an increase of approximately 2 % in classification performance is observed if duplicate pathways are removed. However removing the duplicate pathways does not increase the optimal performance but only increases model performances at lower tolerance values. The increased stability offered by removing duplicate pathways is dramatically observed for entropy criteria models where unexpected performance decrease is completely reversed. Therefore removing the duplicate pathways is a positive pre-processing step that has the effect of removing noise pathways and reducing the effect of any class bias.

Finally we compare the path structure within each model by correlating the

HME3M posterior probabilities for each model classifying each set of extracted pathways. We use the optimal pathways datasets observed in Figure 5 which are; for a scaled expression tolerance of 0, empirical CDF tolerance of 0.5 and an entropy tolerance of -0.66. We then train dataset specific models on a subset of each dataset individually. We use these models classify all pathways extracted from all datasets. Finally we compare the structure within model by correlating each models predicted posterior probabilities with the posterior probabilities of the dataset specific models. The results are presented in Table 1.

Table 1: Correlation between HME3M modeled pathways for each dataset.

Pathway Dataset	HME3M Model					
	Expression With Duplicates	Expression Without Duplicates	Empirical CDF With Duplicates	Empirical CDF Without Duplicates	Expression With Duplicates	Entropy Without Duplicates
Expression With Duplicates	1.000	0.997	0.996	0.971	0.052	0.107
Expression Without Duplicates	0.997	1.000	0.996	0.971	0.045	0.085
Empirical CDF With Duplicates	0.997	0.997	1.000	0.983	0.037	0.109
Empirical CDF Without Duplicates	0.989	0.988	0.987	1.000	0.013	0.069
Entropy With Duplicates	-0.035	0.026	0.028	-0.070	1.000	0.824
Entropy Without Duplicates	0.046	0.090	0.097	-0.078	0.798	1.000

The results in Table 1 show that scaled expression models and empirical CDF models produce highly correlated posterior probabilities when the datasets are switched. This clearly shows that models built from pathways extracted from either scaled expression tolerance or an empirical CDF tolerance will have similar structure. In contrast the entropy models clearly identify a different pathway structure which show little or no correlation with those found by other methods. This difference is due to the low entropy tolerance only extracting paths observed within a single class. Therefore the structure found within the entropy tolerance pathways is unlikely to be representative of the underlying microarray data.

8. Conclusions

In this paper we have compared three criteria for defining an observed pathway through a metabolic network. The results clearly show that employing a rank style transformation such as an empirical CDF upon the raw expressions will extract fewer pathways whilst maintaining the structure within the microarray to produce more stable and accurate pathway classifiers. We have also shown that removing duplicate pathways will increase the stability of the performances and reduce bias. Conversely we have also shown that employing a classification model to extract pathways is likely to produce a severe bias within observed pathway set, leading to unstable results that are not representative of the underlying microarray.

9. Acknowledgments

Timothy Hancock was in part supported by JSPS and in part by BIRD and and Hiroshi Mamitsuka was supported in part by BIRD .

References

[1] Dimitdadou, E., Hornik, K., Leisch, F., Meyer, D., Weingessel, A., e1071 - misc functions of the department of statistics, 2002.

[2] Hancock, T., Mamitsuka, H., A markov classification model for metabolic pathways. *Workshop on Algorithms in Bioinformatics (WABI)*, 2009.

[3] Jordan, M., Jacobs, R., Hierarchical mixtures of experts and the EM algorithm. *Neural Computation*, 6(2):181–214, 1994.

[4] Kanehisa, M., Goto, S., KEGG: Kyoto Encyclopedia of Genes and Genomes. *Nucleic Acids Res.*, 28:27–30, 2000.

[5] Mamitsuka, H., Okuno, Y., Yamaguchi, A., Mining biologically active patterns in metabolic pathways using microarray expression profiles. *SIGKDD Explorations*, 5(2):113–121, 2003.

[6] Park, M.Y., Hastie, T., Penalized logistic regression for detecting gene interactions. *Biostatistics*, 9(1):30-50, 2008.

[7] Schmid, M., Davison, T.S., Henz, S.R., Pape, U.J., Demar, M., Vingron, M., Schölkopf, B., Weigel, D., Lohmann, J.U., A gene expression map of *Arabidopsis thaliana* development. *Nature Genetics*, 37(5):501–506, April 2005.

THE IMPORTANCE OF COMPARTMENTALIZATION IN METABOLIC FLUX MODELS: YEAST AS AN ECOSYSTEM OF ORGANELLES

NIELS KLITGORD[1] DANIEL SEGRÈ[1,2]
niels@bu.edu dsegre@bu.edu

[1] *Graduate Program in Bioinformatics*
[2] *Department of Biology and Department of Biomedical Engineering*
Boston University, Boston MA 02215, U.S.A.

Understanding the evolution and dynamics of metabolism in microbial ecosystems is an ongoing challenge in microbiology. A promising approach towards this goal is the extension of genome-scale flux balance models of metabolism to multiple interacting species. However, since the detailed distribution of metabolic functions among ecosystem members is often unknown, it is important to investigate how compartmentalization of metabolites and reactions affects flux balance predictions. Here, as a first step in this direction, we address the importance of compartmentalization in the well characterized metabolic model of the yeast *Saccharomyces cerevisiae*, which we treat as an "ecosystem of organelles". In addition to addressing the impact that the removal of compartmentalization has on model predictions, we show that by systematically constraining some individual fluxes in a de-compartmentalized version of the model we can significantly reduce the flux prediction errors induced by the removal of compartments. We expect that our analysis will help predict and understand metabolic functions in complex microbial communities. In addition, further study of yeast as an ecosystem of organelles might provide novel insight on the evolution of endosymbiosis and multicellularity.

Keywords: microbial communities; metabolic network models; flux balance analysis; compartments; constraint-based models.

1. Introduction

Microbial ecosystems are ubiquitous on our planet [1], and are implicated in phenomena of utmost importance to humankind: from the global cycling [2] of carbon, nitrogen and sulfur, to the balance between health and disease within our body [3]; from the production of biofuels [4-5] to the evolution of antibiotic resistance [6]. Metabolism, the complex network of biochemical reactions responsible for providing energy and building blocks to the cell, plays a key role in the dynamics and evolution of microbial ecosystems, as most microbe-microbe and microbe-environment interactions are mediated by metabolic intermediates [7-9]. Hence, understanding the flow of metabolism within and between microbes constitutes a fundamental open challenge. While, in terms of computational models, we can confidently predict and understand the metabolic processes of an individual microbial species [10-13], the capacity to predict the concurrent metabolic activity of thousands of different species possibly present in a consortium is currently beyond reach [14]. The magnitude of this problem becomes even more evident upon considering that for most microbial communities there is very little information, if any, about the individual microbial players, their abundance, and their mutual interactions, and that the very definition of species is a debated issue [15]. Metagenomic sequencing data provide incredible panoramic snapshots of the genetic content of microbial communities [16-19], but lack information about how different metabolic reactions are compartmentalized in different species. Thus, deep sequencing technology might provide a list of the metabolic functions performed by the community, a sort of meta-organism, without boundaries between members. Even if it was possible to obtain complete genomic sequences for individual members of a community, genome scale model construction for all organisms, if at all possible, would be an extremely time consuming process.

Paradoxically, it is exactly this lack of information about compartmentalization that may hold the key for a new way of thinking about predictive models of metabolism in microbial ecosystems. As recently suggested by the analysis of multiple microbial consortia [16], the list of metabolic functions encoded in the metagenome may be more predictive of a community environment than the member species composition.

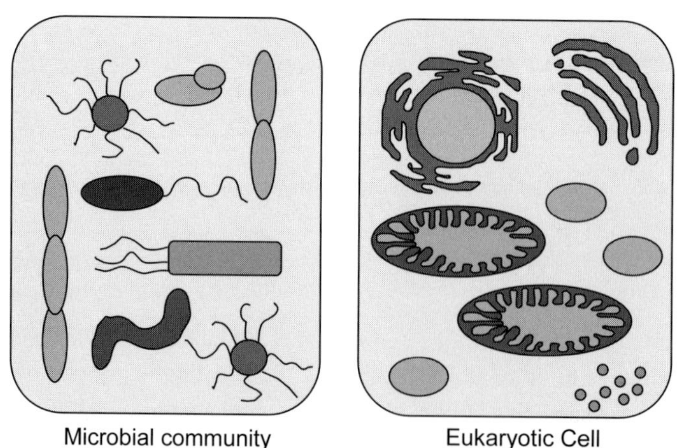

Figure 1: Formal analogy between a microbial ecosystem and a eukaryotic cell, from the perspective of flux balance modeling. Both systems consist of multiple membrane-bound compartments that sequester specific reactions and metabolites. Compartmentalization in flux balance models is implemented by labeling metabolites with the compartment they belong to.

Here, to explore the role of metabolic compartmentalization in microbial ecosystems we focus on the simpler "ecosystem" of organelles in a yeast cell (Figure 1). Eukaryotic cells contain many membrane bound compartments harboring different assortments of metabolic enzymes, and selectively connected by metabolic transport. The yeast flux balance metabolic network model is among the most carefully curated and tested, and it explicitly includes intracellular compartments [20]. Importantly, the way organelles are represented in a eukaryotic metabolic network model is mathematically analogous to the representation of different microbes in an ecosystem. Thus, by studying compartments in the yeast model we expect to gain insight that will be directly useful when modeling a microbial ecosystem. Our work is organized as follows: First, we develop and apply to the yeast model an algebraic method that formally transforms a compartmentalized flux balance model (the c-network) into an elementally balanced de-compartmentalized model (the d-network). To address the impact that the removal of compartmentalization has on model predictions, we next perform a number of metabolic flux comparisons between the c-network and the d-network under different growth media. Finally, we test whether, by systematically constraining some individual fluxes, in the d-network model, to values btained from the c-network solution, we can significantly reduce the flux prediction errors induced by the removal of compartments.

2.   Materials and Method

2.1.   *Constraint-based models of metabolic networks*

Constraint-based models of metabolic networks, such as flux balance analysis (FBA) allow genome-scale simulations of metabolic flow in individual species and have been described in detail elsewhere [21,10]. Briefly, in FBA, a metabolic network is described in terms of an N (reactions) by M (metabolites) stoichiometric matrix S whose element S_{ij} indicates the moles of metabolite i produced (positive sign) or consumed (negative sign) in reaction j (which has flux v_j). Constraints can be applied to the system in the form of a steady state assumption ($Sv=0$), as well as in the form of lower or upper bounds to individual fluxes ($LB_i \leq v_i \leq UB_i$). Within the feasible space defined by these constrains, one can search for reaction flux solutions that maximize or minimize a given (typically linear) objective function. Maximization of the rate of biomass production (representing growth) has been pursued as a standard procedure for FBA in microorganisms. FBA predictions for model organisms such as yeast [13] and *E. coli* [22] have been subjected to extensive experimental validation. Linear Programming calculations were performed using the GNU Linear Programming Kit, and glpkmex for Matlab.

2.2.   *The compartmentalized S. cerevisiae stoichiometric model*

For this analysis we chose to use the recent iMM904 genome scale metabolic reconstruction of the yeast *Saccharomyces cerevisiae* [20]. This model is structured into 8 compartments representing the extracellular space, the cytosol and the yeast organelles: mitochondria, nucleus, golgi, peroxisomes, vacuoles, and endoplasmic reticulum (Table 1). In a stoichiometric model, compartments are usually defined simply by multiple labels associated with metabolites. For example GLC is the label for glucose; in the yeast model, GLC[c] indicates cytosolic glucose, while GLC[m] stands for mitochondrial glucose. In the stoichiometric matrix S for the yeast model, these are effectively represented as two different metabolites, each associated with a row of S. The names of the metabolites form an ordered set (a cell array in Matlab) whose order reflects the corresponding order in the S matrix rows. Transport between compartments is represented by reactions that convert metabolites with different compartment labels into each other. For example, free diffusion of glucose between cytosol and mitochondria would be represented as GLC[c] $\leftrightarrow$ GLC[m]. We refer to this compartmentalized model as the c-network. The de-compartmentalized model is derived from the c-network by applying the de-compartmentalization process described below. We refer to this second model as the d-network.

Table 1: Number of reactions and metabolites in different compartments of the yeast iMM904 metabolic model.

Compartment	Number Reactions	Number Metabolites
Extracellular	333	164
Cytosol	1119	634
Endoplasmic reticulum	32	28
Peroxisome	109	80
Vacuole	30	24
Nucleus	44	40
Mitochondria	306	241
Golgi	13	17

2.3. *De-compartmentalization process*

We define *de-compartmentalization* (Figure 2) as the operation that transforms a compartmentalized stoichiometric model into a corresponding model in which two or more compartments are joined into a single compartment. An example of de-compartmentalization is illustrated in the toy network of Figure 2A, whose compartments are merged to give rise to the network of Figure 2B. Note that in the transformed model, each metabolite is only represented once, and transport and redundant reactions have been removed.

To obtain the yeast de-compartmentalized model (d-network), this process is applied to the original iMM904 model (c-network). Note that while the example of Figure 2 does not include exchange (transport) reactions with the environment, in de-compartmentalizing the yeast model we want to explicitly maintain the distinction between intracellular space and environment, as well as the corresponding transport reactions across the cytosolic membrane. Therefore the d-network contains two metabolic spaces, the extracellular region and the collapsed cellular space.

2.4. *Flux correlation between c-network and d-network*

To measure the flux distance between the c-network and the d-network, we begin by predicting the vector of fluxes v for the c-network model by a using regular FBA (Eq. 1), with maximization of growth rate as an objective function (vector z) [23,13,12]. The initial solution vector $\widetilde{v}^{\,c}$ is then subjected to a secondary optimization, which minimizes the sum of the absolute values of all fluxes. This step reduces the chance that alternative optima will dominate the flux solutions when comparing c-model and d-model (Eq. 2). To find the predicted set fluxes v^d for the d-network model, we then solve a slightly different optimization problem. In order to find how close the d-model can get to the c-model, we will again first find the optimum $\widetilde{v}^{\,d}$ of the d-model with regular FBA, and then use the minimization of metabolic adjustment (MOMA) algorithm [12], to identify,

$$\widetilde{v}^{\,c} = \max \sum_{i=1}^{m} z_i v_i$$
$$s.t. \quad S \cdot v = 0$$
$$LB_j \le v_j \le UB_j$$
(Eq. 1)

$$v^c = \min \sum_{i=1}^{m} |\, v_i \,|$$
$$s.t. \quad S \cdot v = 0$$
$$\sum z_i v_i = \sum z_i \widetilde{v}_i^{\,c}$$
$$LB_j \le v_j \le UB_j$$
(Eq. 2)

$$v^d = \min \sum_{i=1}^{m} |\, v_i - v_i^c \,|$$
$$s.t. \quad S \cdot v = 0$$
$$\sum z_i v_i = \sum z_i \widetilde{v}_i^{\,d}$$
$$LB_j \le v_j \le UB_j$$
(Eq. 3)

among all alternative optima of the d-network, the one with minimal flux distance relative to the solution flux v^c for the c-network (Eq. 3). Finally, the Pearson correlation between the two sets of fluxes v^c and v^d gives us an estimate of the distance between the two flux solutions. If we assume that the fluxes obtained for the c-network are the true ones [23], then this measurement can be interpreted as an estimate of the effect de-compartmentalization has on flux predictions.

2.5. *Metabolite producibility*

To compare the metabolic capabilities of the c-network and d-network, we want to be able to compute the set of metabolites that are producible by a given flux balance model. This can be done by the following steps: (i) we augment the flux balance model with an exchange flux for every metabolite in the model; (ii) for each of these new metabolite exchange reactions, we perform an FBA maximization where the objective flux is the selected exchange reaction. A non-zero flux for this exchange reaction indicates that the corresponding metabolite is producible by the model; (iii) finally, comparison of metabolite producibility of the c-network to that of the d-network requires applying the metabolite transformation matrix T (Figure 2 and Section 3.1) to the vector of producible metabolites for the c-network.

2.6. *Reaction capability and capacity*

The set of reactions that can sustain a nonzero flux in a given flux balance model constitutes what we call the reaction capability of the model. The maximal and minimal flux values through those feasible reactions are said to be the reaction capacities of the model. The reaction capability and capacity are calculated in parallel. For each reaction j in the model we perform two FBA operations, one maximizing v_j and one minimizing it, and hence obtain a range of feasible flux (v_j^{min}, v_j^{max}). These values, defining the capacity of each reaction j, are computed both for the c-network and the d-network, forming two

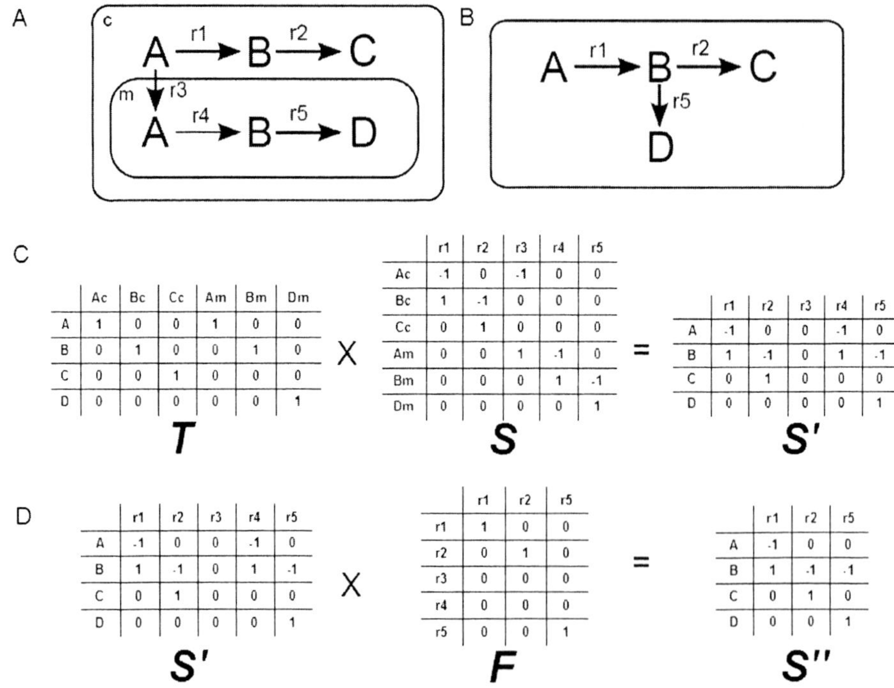

C

T

	Ac	Bc	Cc	Am	Bm	Dm
A	1	0	0	1	0	0
B	0	1	0	0	1	0
C	0	0	1	0	0	0
D	0	0	0	0	0	1

X

S

	r1	r2	r3	r4	r5
Ac	-1	0	-1	0	0
Bc	1	-1	0	0	0
Cc	0	1	0	0	0
Am	0	0	1	-1	0
Bm	0	0	0	1	-1
Dm	0	0	0	0	1

=

S'

	r1	r2	r3	r4	r5
A	-1	0	0	-1	0
B	1	-1	0	1	-1
C	0	1	0	0	0
D	0	0	0	0	1

D

S'

	r1	r2	r3	r4	r5
A	-1	0	0	-1	0
B	1	-1	0	1	-1
C	0	1	0	0	0
D	0	0	0	0	1

X

F

	r1	r2	r5
r1	1	0	0
r2	0	1	0
r3	0	0	0
r4	0	0	0
r5	0	0	1

=

S''

	r1	r2	r5
A	-1	0	0
B	1	-1	-1
C	0	1	0
D	0	0	1

Figure 2: De-compartmentalization algorithm, illustrated in a toy model. The initial network (A) is a two-compartment model. The de-compartmentalized model (B) can be obtained through the transformations shown in C and D, in the two following steps: (i) Build a transformation matrix T that converts into each other the vectors of metabolites before and after the de-compartmentalization. Element T_{ij} is 1 if the new metabolite i in the de-compartmentalized model is the same molecule as metabolite j in the original compartmentalized model (C), and is zero otherwise. This same operator T can then be used to transform the stoichiometric matrix S of the compartmentalized model into the stoichiometric matrix $S'=TS$ of the de-compartmentalized one (C); (ii) Identify all empty rows and all sets of redundant reactions. From this we construct a second transformational matrix F that maps the first of each redundant reaction set to a single location and removes empty reactions (D). By applying this matrix to S' we obtain our de-compartmentalized stoichiometric matrix $S''=S'F$.
To convert the upper bound (UB) and lower bound (LB) vectors into vectors with a number of elements equal to the number of reactions in S'', we employ a variant of matrix F, labeled F' in which all redundant reactions in a set are mapped to a single equivalent reaction. The F' matrix is also used to map between reaction flux solutions from the c-network to the flux reaction space of the d-network (see Methods).

vectors of ranges. Corresponding vectors of reaction capability are defined as having a zero component at position j if $v_j^{min}=v_j^{max}=0$, and having component one otherwise. The c-network vectors are transformed into the d-network vectors by applying the F' matrix (Figure 2). For the purposes of a fair comparison, we only consider reactions for flux capacity when the reaction is capable to sustain nonzero flux in both models.

2.7. *Recovering c-network fluxes in the d-network through individual flux constraints*

First, a flux vector solution v^c is determined for the c-network. This vector is then transformed into the corresponding flux vector for the d-network $v^{c \to d}$ by adding together fluxes in the c-network that correspond to the same reaction (through the F' transformation, Figure 2). We then systematically solve again the FBA model for the d-network (as from Eq. 2), while imposing, for every i, that $v_i^d=v_i^{c \to d}$. For every possible flux constraint added, we evaluate again the Pearson correlation, to determine if that specific constrain is bringing the d-network solution close to the c-network one. This algorithm (which we refer to as "flux-fixing approach") provides us with a list of fluxes whose c-network (i.e. "true") value, when imposed to the d-network, improves substantially the capacity to correctly predict fluxes. Reactions whose flux-constraining operation is able to restore the correlation between d-network and c-network predictions to a value above 0,9 will be called r-reactions. The computation of r-reactions is performed for the 10 different growth media used previously, and shown in Figure 4.

2.8. *Functional enrichment of constraining fluxes*

To determine whether the r-reactions identified across different media tend to belong to specific functional categories, we implemented a functional enrichment analysis. Out of a total of 172 r-reactions identified across all media conditions, 137 are present in all of them. These are r-reactions that can restore the correct fluxes in the d-network with minimal sensitivity to environmental parameters. Hence, we focus on these 137 reactions for our enrichment analysis. The functional annotation for each reaction was taken from the yeast iMM904 model. The significance of enrichment for each pathway was determined using the cumulative hypergeometric distribution function in Matlab. Those pathways that remained significant after an $\alpha=0.05$ false discovery rate correction are reported.

3.   Results and Discussion

3.1. *De-compartmentalization*

Our first step is to convert the original iMM904 model of the yeast network, which includes 8 distinct compartments (Table 1), into a de-compartmentalized version. For this goal we developed an algorithm that removes compartments between reactions, while preserving elemental balance (Figure 2 and Methods). This transformation can be performed through the action of two linear operators on the *S* matrix. The first operator, a metabolite transformation matrix *T*, maps same metabolites from different compartments into single molecular species. The second operator, a refinement matrix *F,* is mostly responsible for collapsing together identical reactions originally belonging to different compartments. The combination of the *T* and *F* operators transforms the c-network model containing 1228 metabolites and 1577 reactions into the d-network model, with 875 metabolites and 1214 reactions. Since the d-network has less variables but also less constraints relative to the c-network, it is not clear *a priori* whether this transformation will contract or expand the space of feasible solutions. In any case, we expect a difference in the specific FBA solutions and in the flux distribution flexibility between c-network and d-network. These differences will be characterized in detail in the next section.

One major effect we might expect as a consequence of the de-compartmentalization process is a disruption of concentration gradients across removed compartment boundaries. This effect may be especially felt in terms of a change in the respiratory capacity (Figure 3). Another consequence we expect is the loss of all inter-compartment metabolite transport costs. This may provide the d-network system with an apparent free energy excess.

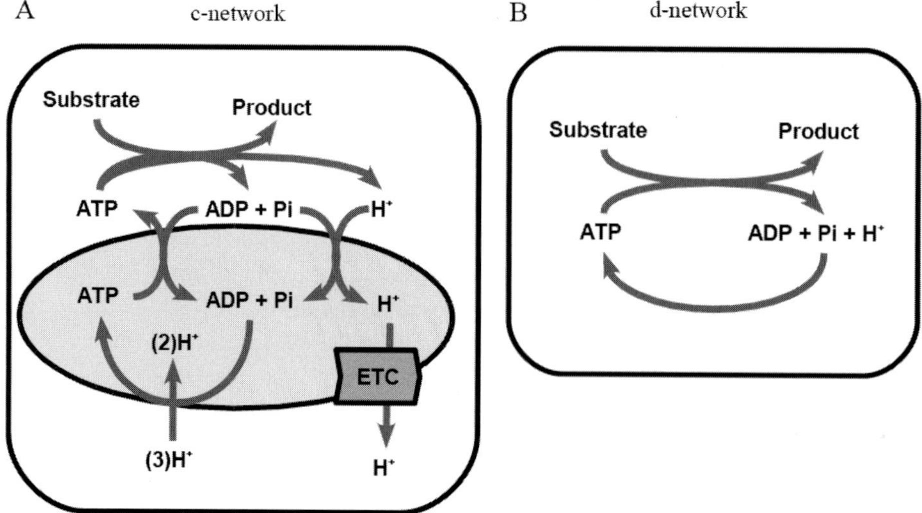

Figure 3: Schematic representation of H$^+$ gradient-driven ATP production in the yeast c-network and d-network. This diagram illustrates the loss of gradient-dependent recharging of ATP upon removal of the mitochondrial membrane. ETC=electron transport chain.

3.2. *Model comparison*

In order to determine whether the d-network model has increased or decreased flexibility in its flux distributions relative to the c-network model, and to ascertain any changes in their predictive ability, we compare the two models using different metrics.

We began by comparing the maximal biomass production (growth) rate between the c-network and the d-network in minimal media. We found that the growth rate for the d-network model is several fold larger than the growth rate for the c-network model. This increase in growth rate was consistently present under 10 different carbon sources, which span a variety of energy potentials, and have entry into the central carbon metabolism at a variety of points. The degree of increase in growth rate depends substantially on the specific carbon used (Table 2).

A more detailed analysis of the extent to which d-network solutions differ from the c-network ones can be performed by comparing the whole set of predicted fluxes for the two models. Specifically, we computed the Pearson correlation coefficient between these two flux solutions under all 10 carbon sources. The correlation coefficients vary between 0.27 and 0.70, depending on the carbon source used (Table 2). Overall, confirming the result obtained by comparing growth rates, this indicates that the FBA predictions for the d-network model display significant discrepancies relative to the "true" c-network model. One may observe that good growth rate agreement does not imply good flux correlation (and vice versa). It is possible to verify that the observed flux discrepancies between c-network and d-network FBA predictions are induced by an increased flexibility of flux capability, flux capacity and metabolite producibility upon de-compartmentalization (see Methods and Table 3).

Table 2: Maximal biomass production rates of the yeast c-network and the d-network, and flux correlation between the two models.

Carbon Source	Uptake Flux (mmol/grDM·h)	Flux Correlation	Growth rate (c-network) (1/h)	Growth rate (d-network) (1/h)
Glucose	10	0.2575	0.2879	1.6333
Fructose	10	0.2721	0.2879	1.6333
Succinate	15	0.4055	0.0914	1.225
Fumerate	15	0.5953	0.0512	1.225
Lactate	20	0.6082	0.0392	1.3611
Pyruvate	20	0.2654	0.0712	1.3611
Acetate	30	0.6981	0.0256	1.2751
Ethanol	30	0.5285	0.0292	1.2751

In order to identify the source of the discrepancies between the c-network and the d-network we have examined in detail the flux distributions for growth on glucose. In agreement with our initial intuition that de-compartmentalization might have a dramatic effects on energy-related pathways, due to the disruption of gradients, we found that the increased biomass yield in the d-network model is caused by an unlimited capacity to recharge ATP from ADP. In the c-network, the ATP synthase is dependent on the differential H^+ concentration gradient between the cytosol and mitochondria. The development of this H^+ gradient is an energy consuming process dependent on reducing NADH or equivalents through the electron transport chain (labeled ETC in Figure 3A). This is not the case in the d-network, where ATP generation is not dependent on the flow of H^+ across the mitochondrial membrane (because it has been removed), but it just relies on the constantly regenerated H^+ in the single intracellular compartment now present (Figure 3B). Thus the ATP recharging reaction (ATP synthase) can carry a constant arbitrarily large flow, as we verified by directly maximizing this flux in the model. Hence energy does not constitute a limiting factor in the d-network, whose biomass production rate will be constrained only by the availability of building blocks (carbon, nitrogen, etc.). It is likely that a large fraction of the discrepancy observed between the c-network and d-network models is ultimately due to this faux unlimited energy supply.

Table 3: Results of metabolic characteristics analysis in the yeast c-network and d-network model. The quantities computed are: metabolite producibility, reaction capability, and reaction capacity, in the c- (see Methods). In all cases the d-network displays more flexibility than the c-network.

Carbon Source	Metabolite Producibility			Reaction Capability			Reaction Capacity		
	c-Network	Both	d-Network	c-Network	Both	d-Network	c-Network	Both	d-Network
Glucose	0	633	6	0	741	95	0	70	671
Fructose	0	633	6	0	741	95	0	70	671
Fumarate	0	632	6	0	739	95	0	62	677
Succinate	0	632	6	0	739	95	0	47	692
Lactate	0	632	6	0	739	95	0	47	692
Pyruvate	0	632	6	0	739	95	0	50	689
Acetate	0	632	6	0	739	95	0	47	692
Ethanol	0	632	6	0	739	95	0	48	691
Citrate	0	632	6	0	739	95	0	55	684
Glycerol	0	632	6	0	739	95	0	51	688

3.3. *Improved de-compartmentalized model through individual flux constraints*

Our analysis to this point suggests that a de-compartmentalized model has increased flux flexibility relative to the corresponding compartmentalized model. This flexibility, due in part to the absence of limitations in energy supply, causes the solutions of the d-network model to be inaccurate, as quantified with flux correlations above. Here we discuss a method to potentially recover this correlation, hence providing a route towards improved de-compartmentalized models. Specifically, we ask whether the flux correlation can be improved by imposing the value of an individual c-network flux as a constraint in the d-network model.

We thus examined how the flux correlation consistency changed when we constrained the d-network FBA model using individual reaction fluxes from the c-network solution, under the 10 media conditions considered. The resulting distribution of correlations between c-network and d-network fluxes shows a clear bimodal distribution (Figure 4). It can be seen that fixing a single flux is sufficient to restore the flux correlation to above 0.9 in 159 cases for growth on glucose. Similar results are obtained for the 9 other media conditions (Table 4). There are a total of 137 reactions that can restore the flux correlation in all 10 media conditions.

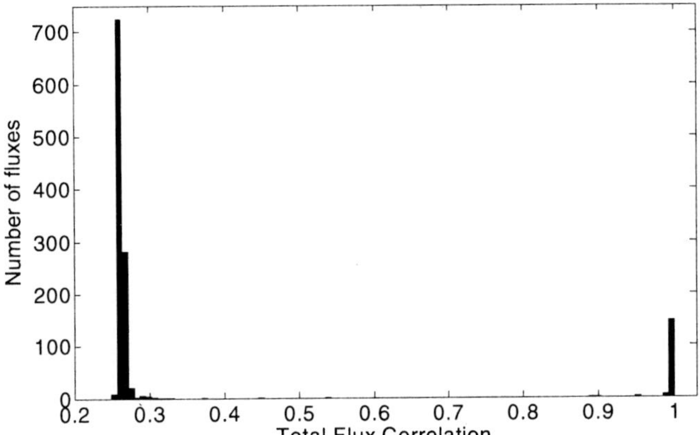

Figure 4: Distribution of flux correlations between the d-network and the c-network upon fixing individual reaction fluxes in a glucose medium. The bi-modal distribution indicates that constrained fluxes have either very little or a very strong effect on the resulting flux correlations.

Visual inspection of these 137 reactions revealed some trivial cases, such as the biomass flux itself, or reactions feeding directly into biomass. Other reactions appeared to be much more interesting, such as those associated with central carbon metabolism or exchange of metabolites with the environment, such as ammonium, sulfate or phosphate.

Table 4: Number of reactions, in the flux-fixing method, that restore a correlation greater than 0.9 between c-network and d-network flux solutions. Numbers are reported for ten different growth media.

Carbon Source	r-reactions	Carbon Source	r-reactions
Glucose	156	Pyruvate	155
Fructose	159	Acetate	158
Fumarate	160	Ethanol	168
Succinate	159	Citrate	156
Lactate	157	Glycerol	148

To assess more comprehensively whether these reactions tend to belong to specific pathways, we performed a functional enrichment analysis (see Methods for details). This revealed an enrichment in central carbon metabolism reactions (Table 5), as well as in numerous pathways associated with amino acid, nucleotide, and lipid metabolism.

The reactions whose c-network flux can be successfully used as a constraint to restore the predictive capacity in the d-network may correspond to critical control points for the biomass production process. Notably, when imposing each of these flux values as constraint, one effectively overcomes the unlimited energy supply problem. This is probably due to the fact that such fluxes are heavily affected by energy limitations in the c-network, and hence "transfer" this information when imposed onto the d-network.

Table 5: Metabolic function enrichment of reactions found to restore d-network fluxes under all 10 media sets assayed. Significance of enrichment was determined by a cumulative hypergeometric function in Matlab. Values reported remained significant after controlling for a 0.05 α False Discovery Rate.

P-value	Pathway
$<10^{-20}$	Valine Leucine and Isoleucine Metabolism
$7.0\text{x}10^{-9}$	Histidine Metabolism
$1.1\text{x}10^{-7}$	Purine and Pyrimidine Biosynthesis
$1.1\text{x}10^{-6}$	Sterol Metabolism
$1.2\text{x}10^{-6}$	Tyrosine Tryptophan and Phenylalanine Metabolism
$7.3\text{x}10^{-6}$	Riboflavin Metabolism
$3.5\text{x}10^{-4}$	Threonine and Lysine Metabolism
$2.0\text{x}10^{-3}$	Fructose and Mannose Metabolism
$2.6\text{x}10^{-3}$	Phospholipid Biosynthesis
$5.2\text{x}10^{-3}$	Glutamine Metabolism

4.   Conclusions

Our comparison of a compartmentalized and a de-compartmentalized yeast metabolism model shows that the spatial separation of metabolites and biochemical reactions in constraint based models is an important feature, especially for a correct assessment of energetic limitations. However, we found that it is possible to restore the predictive capacity of a de-compartmentalized model by constraining certain fluxes to their true (compartmentalized model) values.

We have developed and tested these approaches for a yeast cell flux balance model, thought of as an "ecosystem of organelles", formally analogous to a microbial ecosystem model. This analogy allow us to draw some general conclusions that might be relevant for building models of microbial communities based on metagenomic sequencing data. Future metabolic network reconstructions based on metagenomic sequencing would likely lack the resolution of individual species. Hence, the type of FBA model that one could build from such data corresponds to our de-compartmentalized (or d-network) yeast model. As a consequence of our current analysis, we know that predictions based on such an approximated system may not be accurate, and would likely over-estimate growth rates and fluxes. At the same time, our flux constraint approach shows that, in principle, coupling such a de-compartmentalized model with single carefully selected flux measurements may greatly improve the predictive capacity. It is quite promising that several exchange fluxes can play this role, since these fluxes might be more readily measurable.

Alternatively, the fact that the de-compartmentalized model is considerably affected by the energetic problems caused by gradient removal, suggests that an artificial "meta-compartment" could be used to effectively model concentration gradients in metagenomic-based models.

Finally, while in the current work we have focused on yeast mainly for the purpose of understanding the role of compartmentalization in flux balance models, one could use similar approaches to study, from a metabolic perspective, the origin of eukaryotic cells and multi-compartment systems.

Acknowledgments

This work was partially supported by the NASA Astrobiology Institute and the US Department of Energy.

References

[1] Rothschild, L.J., Mancinelli, R.L., Life in extreme environments, *Nature*, 409: 1092-1101, 2001.

[2] Follows, M.J., Dutkiewicz, S., Grant, S., Chisholm, S.W., Emergent Biogeography of Microbial Communities in a Model Ocean, *Science*, 315:1843-1846, 2007.

[3] Turnbaugh, P.J., Ley, R.E., Hamady, M., Fraser-Liggett, C.M., Knight, R., Gordon, J.I., The Human Microbiome Project, *Nature*, 449: 804-810, 2007.

[4] Warnecke, F., Luginbuhl, P., Ivanova, N., Ghassemian, M., Richardson, T.H., Stege, J.T., Cayouette, M., McHardy, A.C., Djordjevic, G., Aboushadi, N., Sorek, R., Tringe, S.G., Podar, M., Martin, H.G., Kunin, V., Dalevi, D., Madejska, J., Kirton, E., Platt, D., Szeto, E., Salamov, A., Barry, K., Mikhailova, N., Kyrpides, N.C., Matson, E.G., Ottesen, E.A., Zhang, X., Hernandez, M., Murillo, C., Acosta, L.G., Rigoutsos, I., Tamayo, G., Green, B.D., Chang, C., Rubin, E.M., Mathur, E.J., Robertson, D.E., Hugenholtz, P., Leadbetter, J.R., Metagenomic and functional analysis of hindgut microbiota of a wood-feeding higher termite, *Nature*, 450: 560-565, 2007.

[5] Stephanopoulos, G., Challenges in Engineering Microbes for Biofuels Production, *Science*, 315: 801-804, 2007.

[6] Diaz-Torres, M.L., Villedieu, A., Hunt, N., McNab, R., Spratt, D.A., Allan, E., Mullany, P., Wilson, M., Determining the antibiotic resistance potential of the indigenous oral microbiota of humans using a metagenomic approach, *FEMS Microbiology Letters*, 258: 257-262, 2006.

[7] Chait, R., Craney, A., Kishony, R., Antibiotic interactions that select against resistance, *Nature*, 446: 668-671, 2007.

[8]   Kolenbrander, P.E., Andersen, R.N., Blehert, D.S., Egland, P.G., Foster, J.S., Palmer, R.J. Jr., Communication among oral bacteria, *Microbiol. Mol. Biol. Rev.*, 66: 486-505, 2002.

[9]   Woyke, T., Teeling, H., Ivanova, N.N., Huntemann, M., Richter, M., Gloeckner, F.O., Boffelli, D., Anderson, I.J., Barry, K.W., Shapiro, H.J., Szeto, E., Kyrpides, N.C., Mussmann, M., Amann, R., Bergin, C., Ruehland, C., Rubin, E.M., Dubilier, N., Symbiosis insights through metagenomic analysis of a microbial consortium, *Nature*, 443: 950-955, 2006.

[10]   Kauffman, K.J., Prakash, P., Edwards, J.S., Advances in flux balance analysis, *Current Opinion in Biotechnology*, 14: 491-496, 2003.

[11]   Schuetz, R., Kuepfer, L., Sauer, U., Systematic evaluation of objective functions for predicting intracellular fluxes in *Escherichia coli*, *Mol Syst Biol*, 3: 119, 2007.

[12]   Segrè, D., Vitkup, D., Church, G.M., Analysis of optimality in natural and perturbed metabolic networks, *Proceedings of the National Academy of Sciences of the United States of America*, 99: 15112-15117, 2002.

[13]   Snitkin, E.S., Dudley, A.M., Janse, D.M., Wong, K., Church, G.M., Segrè, D., Model-driven analysis of experimentally determined growth phenotypes for 465 yeast gene deletion mutants under 16 different conditions, *Genome Biology*, 9(9):R140, 2008.

[14]   Raes, J., Bork, P., Molecular eco-systems biology: towards an understanding of community function, *Nat Rev Micro*, 6: 693-699, 2008.

[15]   Konstantinidis, K.T., Ramette, A., Tiedje, J.M., The bacterial species definition in the genomic era, *Philosophical Transactions of the Royal Society B: Biological Sciences*, 361: 1929-1940, 2006.

[16]   Dinsdale, E.A., Edwards, R.A., Hall, D., Angly, F., Breitbart, M., Brulc, J.M., Furlan, M., Desnues, C., Haynes, M., Li, L., McDaniel, L., Moran, M.A., Nelson, K.E., Nilsson, C., Olson, R., Paul, J., Brito, B.R., Ruan, Y., Swan, B.K., Stevens, R., Valentine, D.L., Thurber, R.V., Wegley, L., White, B.A., Rohwer, F., Functional metagenomic profiling of nine biomes, *Nature*, 452: 629-632, 2008.

[17]   Gill, S.R., Pop, M., Deboy, R.T., Eckburg, P.B., Turnbaugh, P.J., Samuel, B.S., Gordon, J.I., Relman, D.A., Fraser-Liggett, C.M., Nelson, K.E., Metagenomic Analysis of the Human Distal Gut Microbiome, *Science*, 312: 1355-1359, 2006.

[18]   Tringe, S.G., Rubin, E.M., Metagenomics: DNA sequencing of environmental samples, *Nat Rev Genet*, 6: 805-814, 2005.

[19]   Tyson, G.W., Chapman, J., Hugenholtz, P., Allen, E.E., Ram, R.J., Richardson, P.M., Solovyev, V.V., Rubin, E.M., Rokhsar, D.S., Banfield, J.F., Community structure and metabolism through reconstruction of microbial genomes from the environment, *Nature*, 428: 37-43, 2004.

[20]   Herrgård, M.J., Swainston, N., Dobson, P., Dunn, W.B., Arga, K.Y., Arvas, M., Blüthgen, N., Borger, S., Costenoble, R., Heinemann, M., Hucka, M., Le Novère, N., Li, P., Liebermeister, W., Mo, M.L., Oliveira, A.P., Petranovic, D., Pettifer, S., Simeonidis, E., Smallbone, K., Spasić, I., Weichart, D., Brent, R., Broomhead, D.S., Westerhoff, H.V., Kirdar, B., Penttilä, M., Klipp, E., Palsson, B.Ø., Sauer, U., Oliver, S.G., Mendes, P., Nielsen, J., Kell, D.B., A consensus yeast metabolic network reconstruction obtained from a community approach to systems biology, *Nature Biotechnology*, 26: 1155-1160, 2008.

[21] Lee, J.M., Gianchandani, E.P., Papin, J.A., Flux balance analysis in the era of metabolomics, *Briefings in Bioinformatics*, 7: 140-150, 2006.

[22] Edwards, J.S., Palsson, B.O., Metabolic flux balance analysis and the *in silico* analysis of *Escherichia coli* K-12 gene deletions, *BMC Bioinformatics*, 1: 1., 2000.

[23] Snitkin, E.S. , Segrè, D., Optimality criteria for the prediction of metabolic fluxes in yeast mutants, *Genome Informatics*, 20:123-134, 2008.

A STATE SPACE REPRESENTATION OF VAR MODELS WITH SPARSE LEARNING FOR DYNAMIC GENE NETWORKS

KANAME KOJIMA[1]
kaname@ims.u-tokyo.ac.jp

RUI YAMAGUCHI[1]
ruiy@ims.u-tokyo.ac.jp

SEIYA IMOTO[1]
imoto@ims.u-tokyo.ac.jp

MAI YAMAUCHI[1]
cyowako@ims.u-tokyo.ac.jp

MASAO NAGASAKI[1]
masao@ims.u-tokyo.ac.jp

RYO YOSHIDA[2]
yoshidar@ism.ac.jp

TEPPEI SHIMAMURA[1]
shima@ims.u-tokyo.ac.jp

KAZUKO UENO[1]
uepi@ims.u-tokyo.ac.jp

TOMOYUKI HIGUCHI[2]
higuchi@ism.ac.jp

NORIKO GOTOH[1]
ngotoh@ims.u-tokyo.ac.jp

SATORU MIYANO[1]
miyano@ims.u-tokyo.ac.jp

[1] *Human Genome Center, Institute of Medical Science, University of Tokyo, 4-6-1 Shirokanedai, Minato-ku, Tokyo 108-8639, Japan*
[2] *Institute of Statistical Mathematics, 4-6-7 Minami-Azabu, Minato-ku, Tokyo, 106-8569, Japan*

We propose a state space representation of vector autoregressive model and its sparse learning based on L1 regularization to achieve efficient estimation of dynamic gene networks based on time course microarray data. The proposed method can overcome drawbacks of the vector autoregressive model and state space model; the assumption of equal time interval and lack of separation ability of observation and systems noises in the former method and the assumption of modularity of network structure in the latter method. However, in a simple implementation the proposed model requires the calculation of large inverse matrices in a large number of times during parameter estimation process based on EM algorithm. This limits the applicability of the proposed method to a relatively small gene set. We thus introduce a new calculation technique for EM algorithm that does not require the calculation of inverse matrices. The proposed method is applied to time course microarray data of lung cells treated by stimulating EGF receptors and dosing an anticancer drug, Gefitinib. By comparing the estimated network with the control network estimated using non-treated lung cells, perturbed genes by the anticancer drug could be found, whose up- and down-stream genes in the estimated networks may be related to side effects of the anticancer drug.

Keywords: state space model; VAR model; LASSO; dynamic gene networks.

1. Introduction

For statistical estimation of gene networks from time course microarray data, vector autoregressive (VAR) models have widely been used [4, 6, 8], but have some drawbacks. One is due to the experimental design of time course microarray data. That is, by considering the behaviors of gene expressions after dosing some shock, e.g., drug of interest, it is a usual case that the observed time points are not equally spaced [1, 11]. The assumption of VAR models does not fit this common property of time course microarray data, i.e., VAR models assume equal time intervals in time course data. Also, VAR models do not distinguish observation noise and systems noise in their model equation. To extract reliable information from time course gene expression data, we need to separate observation noise from the gene expressions, otherwise the detection power of the method for gene regulations decreases. To overcome the drawbacks of VAR models, the use of state space model (SSM) was proposed [5, 12]. However, SSM considers the modularity of gene networks to reduce the dimension of the data to achieve stable parameter estimation, the modularity assumption is sometimes not suitable for estimation of gene networks; this decreases prediction power of the network structure when the true network does not have module property.

We propose to combine VAR model and SSM to utilize each advantage and cover each disadvantage and define a new statistical model called VAR-SSM. To use the advantage of VAR model, a simply derived EM algorithm for estimating VAR-SSM requires the computation of inverse matrices whose sizes are equal to the number of genes analyzed. This implies that the applicability of VAR-SSM is limited to small number of genes. Therefore, we also establish a new computational technique for EM algorithm to avoid the computation of the inverse matrices. This technique is essential to increase the effectiveness of VAR-SSM. In many cases, structure learnings of graphical models like Bayesian networks are computationally hard problems [2]. For learning VAR-SSM, we employ a sparse learning strategy based on L1 regularization [10]; this transfers the problem of structure learning into the choice of the value of hyperparameter, which can be optimized based on an information criterion.

The proposed method is used for a case-control study of human lung cells. As case cells, we use normal human lung cells treated by EGF and Gefitinib; the latter is an anticancer drug and known as a selective inhibiter of EGF receptors. As control cells, we use those cells without the treatment. Since Gefinitib is known as a selective inhibiter of EGF receptors, case condition ideally must be the same as control condition. However, gene expression profiles in those conditions are actually different due to the existence of off-targets of Gefitinib. The proposed method is applied for estimating gene regulatory networks from time course data in control and case cells, respectively, and we compare these two estimated networks. By focusing on the hub genes, we could unexpectedly find perturbed genes by Gefitinib, which have different expression profiles between two types of cells in early response time.

2. Methods for Estimating Dynamic Gene Networks

2.1. *Existing Methods*

Vector autoregressive model Given gene expression profile vectors of p genes during T time points $\{\boldsymbol{y}_1, \ldots, \boldsymbol{y}_T\}$, the first order vector autoregressive (VAR(1)) model at time point t is given by:

$$\boldsymbol{y}_t = A\boldsymbol{y}_{t-1} + \boldsymbol{\varepsilon}_t,$$

where A is a $p \times p$ autoregressive coefficient matrix, and ε_t is observation noise at time t which is normally distributed with $N(0, \mathrm{diag}(\sigma_1^2, \ldots, \sigma_p^2))$. The (i, j)th element of A, a_{ij}, indicates a temporal regulation from the jth gene to the ith gene, and thus, if $a_{ij} = 0$, no regulation is considered from the jth gene to the ith gene. By estimating A and checking if a_{ij} is zero for all i, j, regulatory network can be constructed. However, due to the assumption of the model, time points in the time course data must be equally spaced, which is unrealistic in actual biological data.

State space model Unlike VAR model, state space model (SSM) can deal with non equally spaced time course data. In SSM, the set of equally spaced entire T time points are defined as $\mathcal{T}$ and the set of time points where gene expressions are measured or observed are denoted as $\mathcal{T}_{obs}$.

Given gene expression profile vectors of p genes and $|\mathcal{T}_{obs}|$ time points $(\boldsymbol{y}_t)_{t \in \mathcal{T}_{obs}}$, SSM is represented by two equations: system model and observation model. System model is given by

$$\boldsymbol{x}_t = A\boldsymbol{x}_{t-1} + \boldsymbol{w}_t,$$

where $\boldsymbol{x}_t$ is a k dimensional hidden variable vector at time t and A is a $k \times k$ autoregressive coefficient matrix. Observation model is given by

$$\boldsymbol{y}_t = C\boldsymbol{x}_t + \boldsymbol{u}_t, t \in \mathcal{N}_{obs},$$

where C is a $p \times k$ matrix which is called observation matrix. $\boldsymbol{w}_t \sim N(0, R)$ and $\boldsymbol{u}_t \sim N(0, Q)$ are called observation noise and system noise, respectively, where Q is $\mathrm{diag}(q_1, \ldots, q_k)$ and R is $\mathrm{diag}(r_1, \ldots, r_p)$. In SSM, the system model describes the progress of the dynamic system in equally spaced time interval, while the observation model represents the generative process of observation data at intermittent time points, by which non-equally spaced time course data can be properly handled. In addition, although system noise and observation noise are distinguished in SSM, these are merged and cannot be distinguished in VAR model.

State space model as modularity network For estimating gene regulatory networks, the number of genes p is order of 10^2 to 10^4, but the currently available time course consists of < 100 time points. In this case, conventional parameter estimation methods for VAR, e.g., maximum likelihood estimation, may fail due to the overfitting. On the other hand, parameters of SSM can be estimated with a maximum likelihood estimation method by setting k to be much smaller than p. In that case, SSM is used as a dimension-reduction method. In the dimensional reduction, gene expressions $\boldsymbol{y}_t$ are regulated by a few latent factors $\boldsymbol{x}_t$ and the dynamic system is explained by $\boldsymbol{x}_t$. From the biological view point, these latent factors may represent unobserved activities of transcription factors which regulate transcriptions of downstream target genes, and genes regulated by the same latent factor are considered as a transcriptional module. Temporal regulatory relationships among the transcriptional modules are expressed by the system model, and thus the matrix A represents a module network [5, 12]. Genes belonging to a module can be detected by statistical test or other machine learning techniques such as $L1$ penalization or greedy search based on several information criteria. Not only a module network, but also we can estimate a gene regulatory network from the parameters of SSM through a VAR expression of SSM given by

$$R^{-1/2}(\boldsymbol{y}_t - \boldsymbol{u}_t) = D'C'R^{-1}C\Lambda ADR^{-1/2}(\boldsymbol{y}_{t-1} - \boldsymbol{u}_{t-1}) + R^{-1/2}C\boldsymbol{w}_t, \qquad (1)$$

where $D = C'R^{-1}C^{-1}A'R^{-1/2}$. For details of the derivation of the VAR expression, see Hirose *et al.* [5]. However, accuracy of an estimated gene regulatory network probably becomes poor if module structures are not clearly embedded in true one. That property of SSM with dimension reduction may limit its applicability.

Here, Table 1 summarizes advantageous points and drawbacks of gene network estimation using VAR model and SSM, in which check marks indicate the advantageous ones; the both models have different and compensatory characteristics. In the next section, we propose a new probabilistic model that overcomes those drawbacks by combining the two models.

Table 1: Advantageous points and drawbacks of VAR model and SSM.

	Gene Network	Data	System noise and observation noise
VAR	✓	Equal interval	Not distinguished
SSM	Modularity is assumed	✓	✓
Proposed	✓	✓	✓

2.2. *State Space Representation of VAR Model (VAR-SSM)*

To address the drawbacks of VAR model and SSM given in Table 1, we propose a new probabilistic model, state space representation of VAR model (VAR-SSM) and develop its parameter estimation method with L1 regularization. VAR-SSM combines preferable properties of VAR and SSM as shown in Table 1. Like a usual SSM, VAR-SSM is given by the following system model and observation model:

$$\boldsymbol{x}_t = A\boldsymbol{x}_{t-1} + \boldsymbol{w}_t,$$
$$\boldsymbol{y}_t = \boldsymbol{x}_t + \boldsymbol{u}_t, t \in \mathcal{T}_{obs}.$$

A notable difference between usual SSM (i.e., SSM with dimensional-reduction) and VAR-SSM is that matrix C in observation model is set to be an identity matrix. Thus, the hidden variable vectors $\boldsymbol{x}_t$ are p dimensional and the size of A in the system model increases from $k \times k$ to $p \times p$. VAR-SSM can deal with non-equally spaced time course data and distinguish system noise and observation noise. In addition, no modularity is assumed in the network to be estimated. However, due to the increase of the size of A, the estimation of A may fail in a usual manner. We thus consider sparse learning for the estimation of A.

2.3. *Parameter Estimation by EM Algorithm with L1 Regularization*

An L1 regularized log likelihood of VAR-SSM is given by

$$\log \int \prod_{t=2}^{T} \frac{|Q|^{-1/2}}{\sqrt{2\pi}^p} \exp\left\{ -\frac{1}{2}(\boldsymbol{x}_t - A\boldsymbol{x}_{t-1})'Q^{-1}(\boldsymbol{x}_t - A\boldsymbol{x}_{t-1}) \right\}$$

$$\prod_{t \in \mathcal{T}_{obs}} \frac{|R|^{-1/2}}{\sqrt{2\pi}^p} \exp\left\{ -\frac{1}{2}(\boldsymbol{y}_t - \boldsymbol{x}_t)'R^{-1}(\boldsymbol{y}_t - \boldsymbol{x}_t) \right\} d\boldsymbol{x}_1 \ldots d\boldsymbol{x}_T + \sum_{i=1}^{p}\sum_{j=1}^{p} \lambda_i |a_{ij}|,$$

$$(2)$$

where A, Q, and R are parameters to be optimized, and λ_i is the L1 regularization parameter. Since it is difficult to estimate parameters by optimizing Equation (2), we use EM algorithm to estimate the parameters of VAR-SSM model. In EM algorithm, E-step and M-step are repeated to update the values of the parameters so that the log-likelihood converges.

E-step Following expectations are required for the parameter estimation in M-step.

$$E[\boldsymbol{x}_t|\boldsymbol{y}_1,\ldots,\boldsymbol{y}_T]$$
$$E[\boldsymbol{x}_t\boldsymbol{x}_t'|\boldsymbol{y}_1,\ldots,\boldsymbol{y}_T] = Var[\boldsymbol{x}_t|\boldsymbol{y}_1,\ldots,\boldsymbol{y}_T]$$
$$+ E[\boldsymbol{x}_t|\boldsymbol{y}_1,\ldots,\boldsymbol{y}_T]E[\boldsymbol{x}_t|\boldsymbol{y}_1,\ldots,\boldsymbol{y}_T]'$$
$$E[\boldsymbol{x}_t\boldsymbol{x}_{t-1}'|\boldsymbol{y}_1,\ldots,\boldsymbol{y}_T] = Cov[\boldsymbol{x}_t\boldsymbol{x}_{t-1}|\boldsymbol{y}_1,\ldots,\boldsymbol{y}_T]$$
$$+ E[\boldsymbol{x}_t|\boldsymbol{y}_1,\ldots,\boldsymbol{y}_T]E[\boldsymbol{x}_{t-1}|\boldsymbol{y}_1,\ldots,\boldsymbol{y}_T]'. \qquad (3)$$

M-step The elements in A are estimated by solving the following equation:

$$\arg\min_{\boldsymbol{a}_i} \left\{ \boldsymbol{a}_i'U\boldsymbol{a}_i + \boldsymbol{z}'\boldsymbol{a}_i + \lambda_i \sum_{j=1}^{p} |a_{ij}| \right\},$$

where $\boldsymbol{a}_i$ is transpose of the ith row vector of A, i.e., $A = (\boldsymbol{a}_1,\ldots,\boldsymbol{a}_p)'$ and $U = \sum_{t=1}^{T-1} E[\boldsymbol{x}_t\boldsymbol{x}_t'|\boldsymbol{y}_1,\ldots,\boldsymbol{y}_T]$, and $\boldsymbol{z}$ is the ith row vector of $-2\sum_{t=2}^{T} E[\boldsymbol{x}_t\boldsymbol{x}_{t-1}'|\boldsymbol{y}_1,\ldots,\boldsymbol{y}_T]$. The ith diagonal element of Q, q_i can be obtained by

$$q_i = \frac{1}{T-1} \left(\hat{\boldsymbol{a}}_i'U\hat{\boldsymbol{a}}_i + \boldsymbol{z}'\hat{\boldsymbol{a}}_i + v \right),$$

where $\hat{\boldsymbol{a}}_i$ is the estimated $\boldsymbol{a}_i$ in the above equation, and v is the (i,i)th element of $\sum_{t=2}^{T} E[\boldsymbol{x}_t\boldsymbol{x}_t'|\boldsymbol{y}_1,\ldots,\boldsymbol{y}_T]$. Similarly, the ith diagonal element of R, r_i can be obtained by the (i,i) th element of

$$\frac{1}{T} \sum_{t \in \mathcal{T}_{obs}} [\boldsymbol{y}_t\boldsymbol{y}_t' - 2\boldsymbol{y}_t E[\boldsymbol{x}_t|\boldsymbol{y}_1,\ldots,\boldsymbol{y}_T]' + E[\boldsymbol{x}_t\boldsymbol{x}_t'|\boldsymbol{y}_1,\ldots,\boldsymbol{y}_T]] \,.$$

2.3.1. *Kalman Filter for E-step*

In E-step, the expectation values in Equation (3) are calculated by Kalman Filter efficiently through the following three steps: prediction, filtering, and smoothing. Here, $Var[\boldsymbol{x}_t]$ given $\boldsymbol{y}_1,\ldots,\boldsymbol{y}_s$ is denoted by $\Sigma_{\boldsymbol{x}_t|\boldsymbol{y}_{1:s}}$, $E[\boldsymbol{x}_t]$ given $\boldsymbol{y}_1,\ldots,\boldsymbol{y}_s$ is denoted by $\boldsymbol{\mu}_{\boldsymbol{x}_t|\boldsymbol{y}_{1:s}}$, and $Cov[\boldsymbol{x}_t,\boldsymbol{x}_{t-1}]$ given $\boldsymbol{y}_1,\ldots,\boldsymbol{y}_s$ is denoted by $\Sigma_{\boldsymbol{x}_t,\boldsymbol{x}_{t-1}|\boldsymbol{y}_{1:s}}$ Prediction, filtering, and smoothing are calculated by the following recursive formulas:

- Prediction:

$$\Sigma_{\boldsymbol{x}_t|\boldsymbol{y}_{1:t-1}} = A\Sigma_{\boldsymbol{x}_{t-1}|\boldsymbol{y}_{1:t-1}}A' + Q,$$
$$\boldsymbol{\mu}_{\boldsymbol{x}_t|\boldsymbol{y}_{1:t-1}} = A\boldsymbol{\mu}_{\boldsymbol{x}_{t-1}|\boldsymbol{y}_{1:t-1}}.$$

- Filtering:

$$\Sigma_{\boldsymbol{x}_t|\boldsymbol{y}_{1:t}} = (C'R^{-1}C + \Sigma^{-1}_{\boldsymbol{x}_t|\boldsymbol{y}_{1:t-1}})^{-1}, \tag{4}$$
$$\boldsymbol{\mu}_{\boldsymbol{x}_t|\boldsymbol{y}_{1:t}} = \boldsymbol{\mu}_{\boldsymbol{x}_t|\boldsymbol{y}_{1:t-1}} + \Sigma_{\boldsymbol{x}_t|\boldsymbol{y}_{1:t}}C'R^{-1}(\boldsymbol{y}_t - C\boldsymbol{\mu}_{\boldsymbol{x}_t|\boldsymbol{y}_{1:t-1}}).$$

- Smoothing:

$$J_t = \Sigma_{\boldsymbol{x}_t|\boldsymbol{y}_1:\boldsymbol{y}_t}A'\Sigma^{-1}_{\boldsymbol{x}_{t+1}|\boldsymbol{y}_1:\boldsymbol{y}_t}, \tag{5}$$
$$\Sigma_{\boldsymbol{x}_t|\boldsymbol{y}_1:\boldsymbol{y}_T} = \Sigma_{\boldsymbol{x}_t|\boldsymbol{y}_1:\boldsymbol{y}_t} + J_t(\Sigma_{\boldsymbol{x}_{t+1}|\boldsymbol{y}_1:T} - \Sigma_{\boldsymbol{x}_{t+1}|\boldsymbol{y}_1:\boldsymbol{y}_t})J'_{t-1},$$
$$\Sigma_{\boldsymbol{x}_t,\boldsymbol{x}_{t-1}|\boldsymbol{y}_1:\boldsymbol{y}_T} = \Sigma_{\boldsymbol{x}_t|\boldsymbol{y}_1:\boldsymbol{y}_t}J'_{t-1} + J_t(\Sigma_{\boldsymbol{x}_{t+1},\boldsymbol{x}_t|\boldsymbol{y}_1:\boldsymbol{y}_T} - A\Sigma_{\boldsymbol{x}_t|\boldsymbol{y}_1:\boldsymbol{y}_t})J'_{t-1},$$
$$\boldsymbol{\mu}_{\boldsymbol{x}_t|\boldsymbol{y}_1:\boldsymbol{y}_T} = \boldsymbol{\mu}_{\boldsymbol{x}_t|\boldsymbol{y}_1:\boldsymbol{y}_t} + J_t(\boldsymbol{\mu}_{\boldsymbol{x}_t|\boldsymbol{y}_1:\boldsymbol{y}_T} - A\boldsymbol{\mu}_{\boldsymbol{x}_t|\boldsymbol{y}_1:\boldsymbol{y}_t}),$$

where $\Sigma_{\boldsymbol{x}_T,\boldsymbol{x}_{T-1}|\boldsymbol{y}_1:\boldsymbol{y}_T}$ is initialized by

$$\Sigma_{\boldsymbol{x}_T,\boldsymbol{x}_{T-1}|\boldsymbol{y}_1:\boldsymbol{y}_T} = (I - \Sigma_{\boldsymbol{x}_T|\boldsymbol{y}_1:\boldsymbol{y}_T}C^TR^{-1}C)A\Sigma_{\boldsymbol{x}_{T-1}|\boldsymbol{y}_1:\boldsymbol{y}_{T-1}}.$$

Equations (4) and (5) require calculation of $p \times p$ inverse matrix, which is computationally heavy task and numerically unstable as p is order of 10^2 to 10^4. In order to avoid these inverse matrix calculations, we derived a recursive formula in the next section.

2.3.2. *Recursive Formula*

Given $n \times n$ matrix A, $m \times n$ matrix B, and $m \times m$ diagonal matrix $\Delta = \mathrm{diag}(\delta_1,\ldots,\delta_m)$, we consider a matrix formula $(A^{-1} + B^T\Delta B)^{-1}$ for computing. Let $A_0 = A$. By calculating the recursive formula:

$$A_{i+1} = A_i - \frac{1}{1/\delta_{i+1} + \boldsymbol{b}'_{i+1}A_i\boldsymbol{b}_{i+1}}A_i\boldsymbol{b}_{i+1}\boldsymbol{b}'_{i+1}A'_i,$$

where $\boldsymbol{b}_i$ is the ith row vector of B, and $(A^{-1} + B'\Delta B)^{-1}$ is given by A_n.

Proof. From blockwise matrix inversion theorem in Matrix Analysis for Statistics [7] p. 9, we have

$$
\begin{aligned}
A_{i+1} &= A_i - \frac{1}{1/\delta_{i+1} + \boldsymbol{b}'_{i+1} A_i \boldsymbol{b}_{i+1}} A_i \boldsymbol{b}_{i+1} \boldsymbol{b}'_{i+1} A'_i \\
&= (A_i^{-1} + \delta_{i+1} \boldsymbol{b}_{i+1} \boldsymbol{b}'_{i+1})^{-1} \\
&= (A_i^{-1} + \sum_{j=1}^{i+1} \delta_j \boldsymbol{b}_j \boldsymbol{b}'_j)^{-1}.
\end{aligned}
$$

Since $B'\Delta B = \sum_{i=1}^{n} \delta_i \boldsymbol{b}_i \boldsymbol{b}'_i$, $(A^{-1} + B'\Delta B)^{-1}$ is given by A_n. $\square$

Since $\Sigma_{\boldsymbol{x}_{t+1}|\boldsymbol{y}_1:\boldsymbol{y}_t}^{-1}$ in Equation (5) can be given by

$$
\Sigma_{\boldsymbol{x}_{t+1}|\boldsymbol{y}_1:\boldsymbol{y}_t}^{-1} = A'(Q^{-1} - Q^{-1}A(A'Q^{-1}A + \Sigma_{\boldsymbol{x}_t|\boldsymbol{y}_1:\boldsymbol{y}_t}^{-1})^{-1}A'Q^{-1}),
$$

Equations (4) and (5) can be calculated by the recursive formula without inverse matrix calculation.

2.3.3. *L1 Regularized Estimation in M-Step*

In M-step, parameters in A are estimated by minimizing

$$
\arg\min_{\boldsymbol{a}_i} \left\{ \boldsymbol{a}'_i U \boldsymbol{a}_i + \boldsymbol{z}_i \boldsymbol{a}_i + \lambda_i \sum_{j=1}^{p} |a_{ij}| \right\}, \tag{6}
$$

where $A = (\boldsymbol{a}_1, \ldots, \boldsymbol{a}_p)'$ and $U = \sum_{t=1}^{T-1} E[\boldsymbol{x}_t \boldsymbol{x}'_t | \boldsymbol{y}_1, \ldots, \boldsymbol{y}_T]$, and $\boldsymbol{z}_i$ is transpose of the ith row vector of $-2\sum_{t=2}^{T} E[\boldsymbol{x}_t \boldsymbol{x}'_{t-1} | \boldsymbol{y}_1, \ldots, \boldsymbol{y}_T]$. For the better performance of the variable selection, weights $w_1, \ldots, w_p$ are given to the regularization term, which is known as weighted LASSO [9]:

$$
\arg\min_{\boldsymbol{a}_i} \left\{ \boldsymbol{a}'_i U \boldsymbol{a}_i + \boldsymbol{z}' \boldsymbol{a}_i + \lambda_i \sum_{j=1}^{p} w_j |a_{ij}| \right\}. \tag{7}
$$

We set the weight w_j as $\log(|a_{ij}| + 1 + \zeta)$, where ζ is some small value. Note that by using $W = \mathrm{diag}(w_1, \ldots, w_p)$, optimization problem in Equation (7) can be given as

$$
\arg\min_{\boldsymbol{a}_i} \left\{ \boldsymbol{a}'_i W^{-1} U W^{-1} \boldsymbol{a}_i + \boldsymbol{z}' W^{-1} \boldsymbol{a}_i + \lambda_i \sum |a_{ij}| \right\}. \tag{8}
$$

Since the form of Equation (8) is the same as that of Equation (6), for brevity, optimization of Equation (6) is shown in the next section.

2.3.4. *LARS-LASSO for VAR-SSM*

In a linear regression problem $y = X\beta + \varepsilon$, where y is the vector of responses, X is the design matrix and ε is the noise. For the estimation of β, LASSO [10] solves the following optimization problem:

$$\arg\min_{\boldsymbol{\beta}} \left\{ (\boldsymbol{y} - X\boldsymbol{\beta})'(\boldsymbol{y} - X\boldsymbol{\beta}) + \lambda_i \sum_i |\beta_i| \right\}. \tag{9}$$

The exact solution of Equation (9) can be obtained by the LARS-LASSO algorithm [3] in an efficient manner, but the LARS-LASSO algorithm is limited to the equation given by Equation (9) and thus cannot solve the optimization of Equation (6) for VAR-SSM. Thus, we derived a new LARS-LASSO algorithm for M-step of VAR-SSM. We first define a function $f(\boldsymbol{a}_i)$ as

$$f(\boldsymbol{a}_i) = \boldsymbol{a}_i' U \boldsymbol{a}_i + \boldsymbol{z}' \boldsymbol{a}_i + \lambda_i \sum_j |a_{ij}|.$$

Let $\mathcal{A}$ be an active set of indices, i.e., the set of indices whose corresponding coefficient is not zero. From Karush-Kuhn-Tucker conditions, we have

$$\frac{d}{d\boldsymbol{a}_{i,\mathcal{A}}} f(\boldsymbol{a}_i) = 2U_{\mathcal{A}} \boldsymbol{a}_{i,\mathcal{A}} + \boldsymbol{z}_{\mathcal{A}} + \lambda_i \boldsymbol{s}_{\mathcal{A}} = \boldsymbol{0}, \tag{10}$$

where $\boldsymbol{a}_{i,\mathcal{A}} = (a_{ij})'_{j\in\mathcal{A}}$, $\boldsymbol{z}_{\mathcal{A}} = (z_j)'_{j\in\mathcal{A}}$, $U_{\mathcal{A}} = (u_{jk})_{j,k\in\mathcal{A}}$, and $\boldsymbol{s}_{\mathcal{A}} = (\mathrm{sgn}(a_{ij}))'_{j\in\mathcal{A}}$, and thus $\boldsymbol{a}_{i,\mathcal{A}}$ can be given by

$$\boldsymbol{a}_{i,\mathcal{A}} = -\frac{1}{2} U_{\mathcal{A}}^{-1} \left(\boldsymbol{z}_{\mathcal{A}} + \lambda_i \boldsymbol{s}_{\mathcal{A}} \right).$$

As far as $\mathcal{A}$ does not change, Equation (10) holds if λ changes continuously. Thus, if λ_i decreases by γ and $\mathcal{A}$ does not change, $\boldsymbol{a}_i$ is given by

$$\boldsymbol{a}_{i,\mathcal{A}}(\lambda_i + \gamma) = \boldsymbol{a}_{i,\mathcal{A}}(\lambda_i) + \frac{\gamma}{2} U_{\mathcal{A}}^{-1} \boldsymbol{s}_{\mathcal{A}}. \tag{11}$$

Thus, when decreasing λ_i from λ_{max}, we need to focus on the point where $\mathcal{A}$ changes. We suppose that $\mathcal{A}$ changes at $\tilde{\lambda}_j$. From Equation (10), the following elementwise equation also holds:

$$|2U_{\mathcal{A}} \boldsymbol{a}_{i,\mathcal{A}} + \boldsymbol{z}_{\mathcal{A}}| = \lambda_i. \tag{12}$$

If the index $j \notin \mathcal{A}$ is added when $\tilde{\lambda}_j$ decreases by γ_j, from Equation (12), $|z_j| = \tilde{\lambda}_j - \gamma_j$ holds. On the other hand, if the index $j \in \mathcal{A}$ is deleted, i.e., $a_j(\tilde{\lambda}_i - \gamma_j) = 0$, from Equation (11), $2a_{ij}/o_j = \gamma_j$ holds, where o_j is an element of $(1/2) \cdot U_{\mathcal{A}}^{-1} \boldsymbol{s}_{\mathcal{A}}$ corresponding to the index j. Since $\mathcal{A}$ actually changes at the closest point satisfying the above cases to $\tilde{\lambda}_i$, the point where $\mathcal{A}$ changes is $\tilde{\lambda}_i - \gamma$, where γ is defined by

$$\gamma = \min\{\gamma_{na}, \gamma_a\},$$

where γ_{na} is given by

$$\gamma_{na} = \min_{j\notin\mathcal{A}} \gamma_j > 0 \quad \text{s.t.} \quad |z_j| = \tilde{\lambda}_i - \gamma_j,$$

and γ_a is given by

$$\gamma_a = \min_{j\in\mathcal{A}} \frac{2a_{ij}}{o_j} > 0.$$

From Equation (12), the maximum value $\tilde{\lambda}_i$ can take is given by $\max_j |z_j|$ when $\mathcal{A} = \emptyset$. Hence, beginning from $\mathcal{A} = \emptyset$ and $\tilde{\lambda}_i = \max_j |z_j|$, we can obtain $\boldsymbol{a}_i$ by computing the above procedure until $\tilde{\lambda}_i$ reaches some specified λ_i.

3. Numerical Experiments

We apply VAR-SSM, VAR model, and SSM with dimension reduction to simulated datasets to evaluate performances of the proposed method. To generate simulation datasets, a directed scale free network of 100 nodes and 150 edges is generated and self regulation edges are added to root nodes of the network. Finally, we obtain the network of 181 edges which includes 31 self regulation edges. The AR coefficients are uniformly assigned to the edges from $\{-0.9, -0.8, -0.7, -0.6, -0.5, 0.5, 0.6, 0.7, 0.8, 0.9\}$. System noise $\boldsymbol{w}_t$ and observation noise $\boldsymbol{u}_t$ are set to follow $N(0, I)$ and $N(0, 0.2I)$, respectively. In the following experiments, the three network estimation methods are applied to 100 datasets generated from the above network and averaged results are used for comparison.

3.1. *Comparison to VAR*

We compare performances of VAR-SSM and VAR model by applying to simulated datasets in which non-equally spaced time course data are generated by the following procedures:

- Prepare equally spaced time course data of 100 time points from the scale free network model.
- The first 40 points are directly used, every two points is left in the next 30 points, and every three point is left in the final 30 points. Thus, no-equally spaced time course data of 65 time points are used.

The numbers of true positives, false positives, true negatives, and false negatives for estimated networks by VAR model and VAR-SSM are summarized in Table 2. Since the elements of the coefficient $p \times p$ matrix in VAR model should be estimated and determined whether zero or not, like VAR-SSM we use LASSO for sparse learning.

Table 2: The numbers of true positives (# TP), false positives (# FP), true negatives (# TN), and false negatives (# FN) of VAR and VAR-SSM.

	# TP	# FP	# TN	# FN
VAR (LASSO)	138	1851	43	7968
VAR-SSM	111	47	70	9772

3.2. *Comparison to Usual SSM*

We next compare the performances of VAR-SSM and SSM with dimension reduction. To estimate a gene regulatory network from usual SSM, we utilize the VAR representation of usual SSM given by Equation (1). Although Hirose *et al.* [5] applied permutation test to detect significant edges in gene regulatory networks, it is computationally heavy task. Thus, instead of the permutation test, we select edges by a threshold value for absolute values of the estimated AR coefficients in Equation (1). Thus, by changing the threshold value, ROC like curves can be drawn, in which vertical axis indicates the number of true positive edges, while horizontal axis indicates the number of false positive edges.

Figure 1 gives the ROC like curves of SSM for time course data of 40, 60, 100 time points. A point represented by cross in each ROC like curve indicates the results of VAR-SSM. Since, in the ROC like curves, left upper area indicates better performance, the results suggest that VAR-SSM is better than usual SSM. In addition, for longer time course data, false positives of VAR-SSM are drastically reduced, while its true positives seem saturated and do not change a lot. Also the results of SSM are essentially unchanged if the number of time points increases. We guess the reason is that the performance of SSM was saturated when 40 time points were used. This saturation might be due to the true network structure; the true network structure does not follow the modularity assumption in SSM.

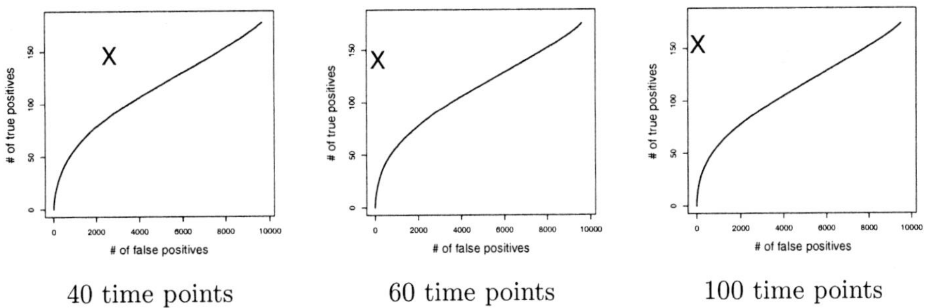

40 time points   60 time points   100 time points

Fig. 1: ROC like curves of usual SSM for time course data of 40, 60, and 100 time points. The numbers of true positives and false positives of VAR-SSM are indicated by cross

3.3. *Real Data Application*

We apply VAR-SSM to time course gene expression data of normal human Small Airway Epithelial Cells (SAECs), which are lung cells. In lung cancer cells, EGF receptors (EGFRs) are often overexpressed. By stimulating EGFRs in SAEC with EGF, lung cancer cells might be simulated. An anticancer drug, Gefitinib is known as a selective inhibiter of EGFRs. Thus we can expect that SAECs dosed with both EGF and Gefitinib are ideally in the same state of SAECs without the stimulation. However, we observed that some gene expression patterns were different between the two conditions. These genes may represent unknown action mechanisms of Gefitinib and relate to side effects. Gene regulatory networks constructed from these data probably give some insights about the above points. We, thus, employ SAECs dosed with EGF and Gefitinib as case cells and SAECs without the stimulation as control cells and consider a case-control study based on network viewpoint.

We first estimate gene regulatory networks by applying VAR-SSM to time course gene expression data of the case and those of the control for 100 genes, respectively. These 100 genes are selected as follows:

- 500 genes with large variation coefficients are learnt by SSM.
- Well learnt 100 genes with respect to p-value defined in [11] are selected.

The time course data consist of 19 time points during 48 hours and are not equally spaced. We next focus on hub genes in estimated networks and analyze how many edges connected to each of the genes in both networks. Hereafter, the number of edges connected to a node is referred to as degree. We calculate how much degree from the control network to the case network increases for each of the genes and rank the genes with respect to the increased degree. Top six genes in the ranking are listed in Table 3. Differences of expression profiles of the top genes between the case and control cells are shown in Figure 2. Many of the top genes are differentially expressed between control cell and case cell in the beginning stage and are considered as perturbed genes by Gefitinib. We consider that genes around these perturbed genes in the estimated case network may be off-targets of Gefitinib and related to the side effects of Gefitinib.

Table 3: Ranking of genes with respect to increased degree

Gene	Degree in Control	Degree in Case	Increased Degree
FOS	3	12	9
SOCS3	2	10	8
LRFN1	21	29	7
KLF2	4	11	7
SNX26	2	9	7
BMF	3	9	6

4. Conclusion

In this study, we proposed a new probabilistic model, VAR-SSM and showed its parameter estimation method using a sparse learning technique. Efficient calculation technique for recursive formula in E-step and a new LARS algorithm for M-step enable the parameter estimation. Simulation studies using time course data from a

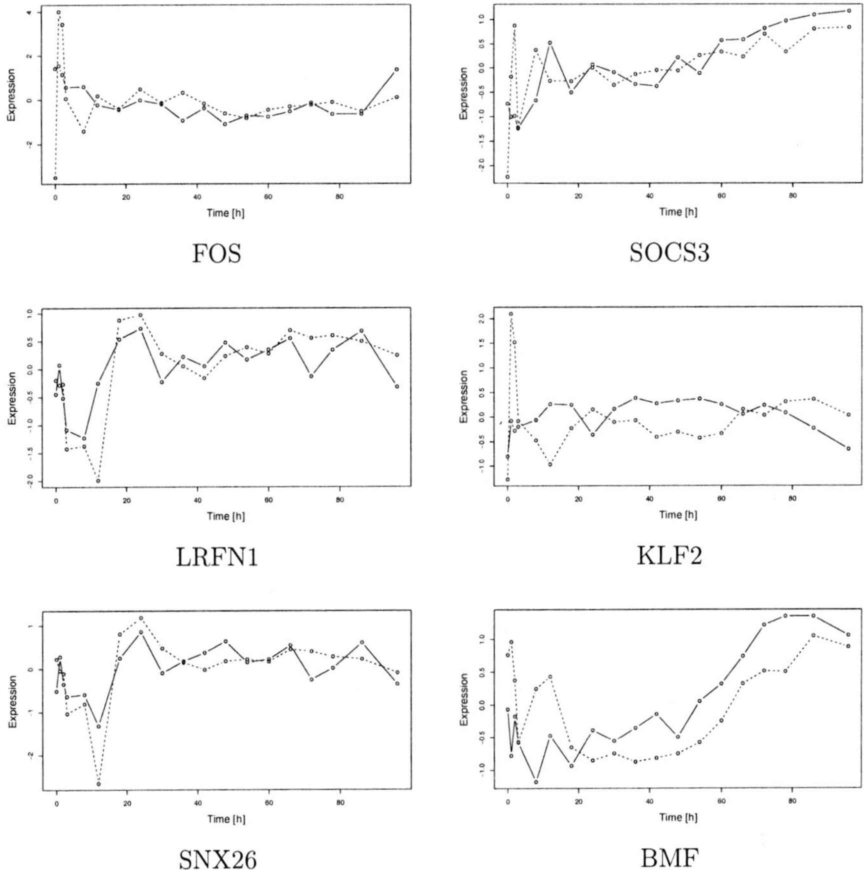

FOS

SOCS3

LRFN1

KLF2

SNX26

BMF

Fig. 2: Expression profiles of top genes in ranking given by Table 3. Expression profiles from control cell and case cell are indicated by solid and dashed lines, respectively.

scale free network shows that the proposed model outperforms VAR model and SSM with dimension reduction. For a real data application, VAR-SSM is used to estimate gene regulatory networks using time course data from human normal lung cells treated by EGF and Gefitinib and non-treated cells. From the estimated networks, we could find candidates of perturbed genes by Gefitinib. By investigating genes around candidate perturbed genes in the estimated networks, we would like to elucidate factors of the side effect of Gefitinib and their mechanisms in future work.

References

[1] Bar-Joseph, Z., Gerber, G., Simon, I., Gifford, D., Jaakkola, T., Comparing the continuous representation of time-serise expression profiles to identify

differentially expressed genes, *Proceedings of the National Academy of Science*, 100(18):10146–10151, 2003.

[2] Chickering, D.M., Heckerman, D., Meek, C., Large-sample learning of Bayesian neworks is NP-hard. *Journal of Machine Learning Research*, 15:1287–1330, 2004.

[3] Efron, B., Hastie, T., Johnstone, J., Tibshirani, R., Least angle regression, *Annals of Statistics*, 32(2):407–499, 2004.

[4] Fujita, A., Sato, J.R., Garay-Malpartida, H.M., Morettin, P.A., Yamaguchi, R., Miyano, S., Sogayar, M.C., Ferreira, C.E., Modeling gene expression regulatory networks with the sparse vector autoregressive model, *BMC Systems Biology*, 1:39, 2007.

[5] Hirose, O., Yoshida, R., Imoto, S., Yamaguchi, R., Higuchi, T., Stephen, D., Chamock-Jones, C., Miyano, S., Statistical inference of transcriptional module-based gene networks fro time course gene expression profiles by using state space models, *Bioinformatics*, 24(7):932–942, 2008.

[6] Kojima, K., Fujita, A., Shimamura, T., Imoto, S., Miyano, S., Estimation of nonlinear gene regulatory networks via L1 regularized NVAR from time series gene expression data, *Genome Informatics*, 20:37–51, 2008.

[7] Schott, J.R., *Matrix Analysis for Statistics*, John Wiley & Sons, Inc., 1997.

[8] Shimamura, T., Imoto, S., Yamaguchi, R., Fujita, A., Nagasaki, M., Miyano, S., Recursive elastic net for inferring large-scale gene networks from time course microarray data, *BMC Systems Biology*, 3:41, 2009.

[9] Shimamura, T., Imoto, S., Yamaguchi, R., Miyano, S., Weighted lasso in graphical Gaussian modeling for large gene network estimation based on microarray data, *Genome Informatics*, 19:142-153, 2007.

[10] Tibshirani, R., Regression shrinkage and selection via the lasso, *Journal of Royal Statistical Scociety: B*, 5(1):267–288, 1996.

[11] Yamaguchi, R. Imoto, S., Yamauchi, M., Nagasaki, M, Yoshida, R., Shimamura, T., Hatanaka, Y., Ueno, K., Higuchi, T., Gotoh, N., Miyano, S., Predicting differences in gene regulatory systems by state space models, *Genome Informatics*, 21:101-113, 2008.

[12] Yamaguchi, R. Yoshida, R., Imoto, S. Higuchi, T. Miyano, S., Finding module-based gene networks with state-space models-Mining high-dimensional and short time-course gene expression data, *IEEE Signal Processing Magazine*, 24(1):37-46, 2007.

FORMAL REPRESENTATION OF
THE HIGH OSMOLARITY GLYCEROL PATHWAY IN YEAST

CLEMENS KÜHN[1]

clemens.kuehn@biologie.hu-berlin.de

EDDA KLIPP[1]

edda.klipp@rz.hu-berlin.de

K. V. S PRASAD[2]

prasad@chalmers.se

PETER GENNEMARK[3,4]

peterg@chalmers.se

[1] *Humboldt University, DE-10115 Berlin, Germany.*
[2] *Chalmers University of Technology, SE-412 96 Göteborg, Sweden*
[3] *University of Göteborg, SE-412 96 Göteborg, Sweden*
[4] *Uppsala University, SE-751 06 Uppsala, Sweden*

The high osmolarity glycerol (HOG) signalling system in yeast belongs to the class of Mitogen Activated Protein Kinase (MAPK) pathways that are found in all eukaryotic organisms. It includes at least three scaffold proteins that form complexes, and involves reactions that are strictly dependent on the set of species bound to a certain complex. The scaffold proteins lead to a combinatorial increase in the number of possible states. To date, representations of the HOG pathway have used simplifying assumptions to avoid this combinatorial problem. Such assumptions are hard to make and may obscure or remove essential properties of the system.

This paper presents a detailed generic formal representation of the HOG system without such assumptions, showing the molecular interactions known from the literature. The model takes complexes into account, and summarises existing knowledge in an unambiguous and detailed representation. It can thus be used to anchor discussions about the HOG system. In the commonly used Systems Biology Markup Language (SBML), such a model would need to explicitly enumerate all state variables. The *Kappa* modelling language which we use supports representation of complexes without such enumeration.

To conclude, we compare *Kappa* with a few other modelling languages and software tools that could also be used to represent and model the HOG system.

Keywords: High osmolarity glycerol pathway; rule-based modelling; Kappa.

1. Introduction

The High Osmolarity Glycerol (HOG) signalling pathway in the yeast *Saccharomyces cerevisiae* belongs to the class of Mitogen Activated Protein Kinase (MAPK) pathways found in all eukaryotic organisms, and is of general interest. To unambigouosly represent and communicate current hypotheses of the molecular interactions between the signalling proteins in the HOG system is of fundamental interest. This is a challenging problem for several reasons.

First, many details of the molecular interactions are unknown, and alternative hypotheses with various levels of experimental suppport are present in the literature. The main reason for lack of knowledge is that many phosphorylation states or protein interactions cannot be experimentally observed. We note in passing that the rapid development of measurement techniques like mass spectrometry has great potential to cure this problem and allow simultaneous observation of multiple species [6].

Second, the system includes complexes and many different phosphorylation states, and an exhaustive formal representation therefore requires a large number of state variables, because of the combinatorial increase of possible combinations. For example, the scaffold protein Pbs2 in yeast has at least three docking sites and can, besides, be single and double phosphorylated (Fig. 1). Scaffold proteins can be thought of as hubs in signalling networks [24, 34, 37, 42]: proteins binding to the docking sites can, in turn, have their own phosphorylation state and additional docking sites. A full representation of Pbs2 and its known binding partners (Sho1, Ssk2/22 and Hog1) and their binding partners (Ste11-Ste50 and Ssk1), requires as many as 1900 state variables.

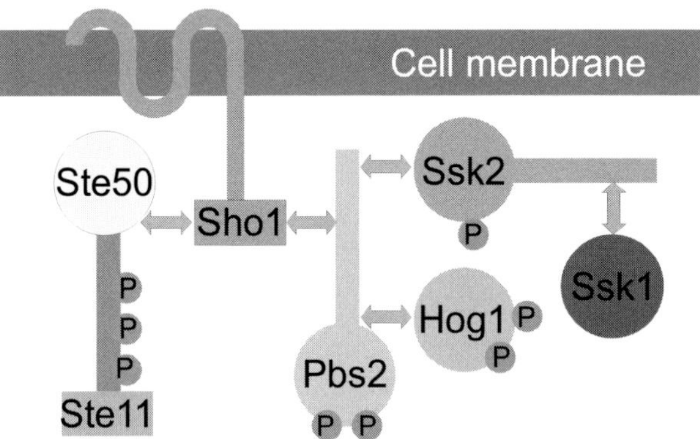

Fig. 1: The yeast scaffold protein Pbs2 has three docking sites and can be phosphorylated on two sites [26, 33, 45]. Double-headed arrows indicate a potential docking of a protein. Small circles denote a potential phosphorylation. To illustrate the combinatorial problem: assuming that all phosphorylation reactions are independent results in 8 states for Ste11-Ste50, 4 states for Pbs2 and for Hog1, and 2 states for Ssk2/22. Besides, some proteins may be absent and Sho1 can be in an inactive or active state. Therefore, in total, 1900 states are possible.

The traditional way to represent the system in journal publications is by an illustration like Fig. 1 and accompanying text in natural language. The text is often extensive and hard to parse and analyse. Representations in this style can therefore be ambiguous.

It is therefore natural to consider more formal approaches. Several mathematical models in the form of ordinary differenrial equations (ODEs) have been proposed. Using ODEs, we can in principle represent the system unambiguously at the level of molecular interactions. However, the combinatorial problem typically leads to a very large model as examplified in Fig. 1. Therefore, in practice, reasonable biological assumptions are used to drastically reduce the number of state variables [4, 5, 7].

For example, for the Pbs2 system in Fig. 1, we can assume that Ssk2 is constitutively bound to Pbs2 as indicated in Ref. [45], and that proteins are either fully phosphorylated or not phosphorylated at all. In some cases, depending on the range of experimental scenarios we want to study, assumptions of this kind can be satisfactory, especially since the amount of information also decreases. However, in many cases this solution is unsatisfactory because essential properties of the system may be obscured or removed, and common system modifications (like *pbs2*Δ and *ssk2*Δ in the above example) cannot be explicitly represented in the model. It is also the case that assumptions that seem plausible for a certain experimental scenario, may not be realistic for other system modifications and/or input functions that are tried later.

We want therefore to represent the relatively well characterised HOG system by a formal model that takes complexes into account, but without explicitly enumerating all states. Several publications in recent years have suggested clues to the detailed interaction structure of the pathway, but only in the form of figures like Fig. 1 and accompanying text. There is hence a need for a formal unambiguous model of the system, allowing efficient exchange of hypotheses and results. Ideally, such models can subsequently be used for simulation and prediction of experiments, as well as in experimental planning to assist the experimentalist.

There are several different modelling languages that support modelling of complexes without enumerating all states, and some of them even represent the complexes explicitly [2, 3, 9, 28]. There are also corresponding software tools. It is notable that the systems biology mark-up language (SBML) currently lacks support for representing complexes in this respect [20]. However, such support is listed as a possible future enhancement of SBML, highlighting the importance of this issue [21].

We set out the following requirements for our model. It should

- unambiguously summarise current knowledge about the system.
- take complexes into account without enumerating all states.
- be easy for biologists to edit, and to use in discussing the system.
- include dynamic information, allowing simulation when data becomes richer.

Because only a few state variables in the system can be measured, the kinetic rate constants of the system are currently not observable. The problem with unidentifiable parameters is common for these kind of systems, and it will not affect the way we *represent* the structure of the system. However, a future goal is naturally to infer a dynamic model of the HOG system.

In this paper, we propose a generic model of the HOG signalling system based on literature, and an implementation of this model in the *Kappa* modelling language [8, 9] using the software tool Cellucidate (http://www.cellucidate.com/). Cellucidate was chosen because it allows all four requirements above to be fulfilled.

Other strong alternatives that we considered were BioNetGen [2] and

PottersWheel [28]. In the Discussion section, we discuss advantages and limitations of these languages and of another relevant language, SPiM [3].

2. Modelling the HOG Pathway

The HOG pathway consists of a sensing system, a cascade of protein kinases and output systems such as transcriptional regulators [18]. To give an overview, a static pathway diagram is given in Fig. 2. Here, some causality has been incorporated by re-drawing certain complexes in series and indicating molecule bindings by arrows. As we said, the traditional way to depict models in the biological literature is through figures of this kind, with ambiguities carefully explained in the text.

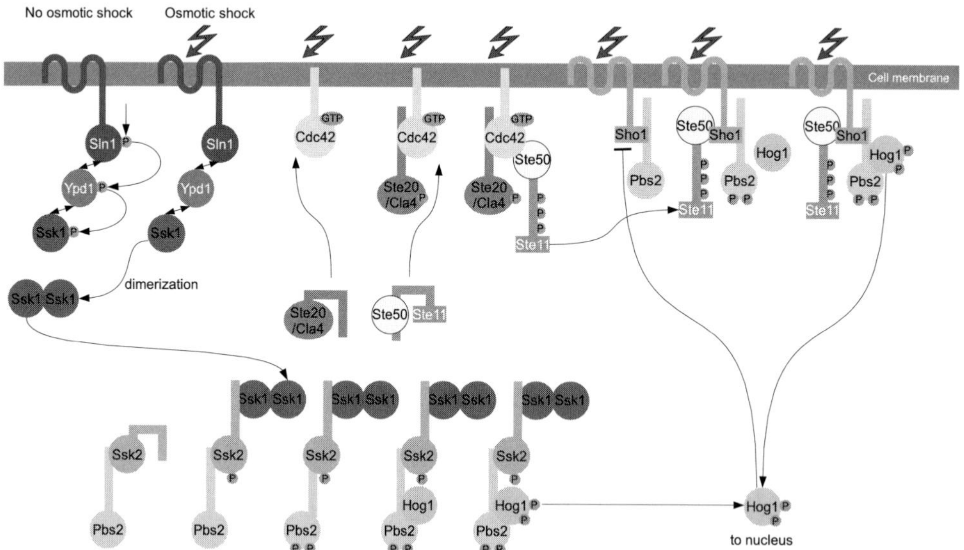

Fig. 2: Traditional model of a signalling pathway reconstructed from the literature. Details and references are given in the text and in Table 1.

The HOG system is activated by osmotic shock, e.g. by the addition of salt to the medium. At least three membrane proteins, Cdc42, Sho1 and Sln1, are activated by the stress, presumably by changes in turgor pressure [40, 44]. Two different branches, the Cdc42-Sho1 branch and the Sln1 branch, merge at the protein Pbs2, which in turn activates Hog1, the last kinase in the pathway. Hog1 activates cytosolic effectors (e.g. Pfk27 [11]) and enters the nucleus and induces transcription. This, in turn, leads to intracellular glycerol accumulation, that counters effects of the osmotic shock by increasing intra-cellular osmotic pressure.

Presently, three scaffold proteins are identified; Cdc42, Sho1 and Pbs2. Some phosphorylation reactions are known to be strictly dependent on the set of species bound to a certain complex. For example,

- the phosphorylation of Ste11 is assumed to require that both Ste50-Ste11 and Ste20 are bound to Cdc42, and besides, that Ste20 is phosphorylated [46].
- the phosphorylation of Pbs2 by Ste11 is assumed to require that both Ste11 and Pbs2 are bound to Sho1, and besides, that Ste11 is phosphorylated [46].
- the phosphorylation of Pbs2 by Ssk2 is assumed to require that unphosphorylated Ssk1 (potentially as a dimer) binds Ssk2 when Ssk2 is in complex with Pbs2 [45].

The HOG system has previously been characterised by mathematical models at various levels of detail [13, 15, 16, 22, 23, 29]. Of these, the most detailed model is that of Klipp *et al.* in Ref. [22].

In Table 1, we propose a generic model of the HOG signalling system built from information in the recent literature, and experimentally supported. Our generic model differs from that of Klipp *et al.* in three main respects:

- We include all known signalling branches.
- We model complexes.
- We focus on the signalling pathways. Gene regulation, metabolism and biophysics of osmoregulation are not included.

At this stage, we have chosen not to take activation of phosphatases and localisation of $Hog1^{PP}$ into account. We could also not yet account for the implications made in the very recent literature [25]. Table 1 should be read in conjunction with Fig. 2.

In Table 1, states of molecules (phosphorylated: X^P, unphosphorylated: X^U) are only specified if required to unambiguously define a particular reaction. To allow simulation, we define an input variable called *Osmostress* representing the biophysical changes that lead to activation of Sln1, Sho1 and Cdc42. Adaptation is mainly driven by $Hog1^{PP}$ dependent accumulation of intracellular glycerol, here modelled by a negative feedback of $Hog1^{PP}$ on *Osmostress*.

3. Implementing the HOG pathway in Kappa

In this section we describe how we implemented the generic HOG model of Table 1 in Kappa, using the web-based application, Cellucidate, and the corresponding stand-alone program, Kappa Factory [8, 9]. The Kappa code is listed in the Supplementary material and on the model repository of Cellucidate's web site (http://www.cellucidate.com).

Kappa is a process calculus in the tradition started by the Calculus of Communicating Systems, referred to as CCS [30]. CCS allows description and analysis of concurrent systems. In Cellucidate a rule-based representation is used, and state variables are not explicitly enumerated [8, 9]. Instead, the model is fully represented by a list of agents (e.g. proteins) and their individual docking sites and activation (e.g. phosphorylation) sites, and a list of rules corresponding to a generalised form of a chemical reaction. Basically, the rules are defined on the same

Table 1: Generic Model of the HOG Pathway.

Reaction	Comment and references
Cdc42 $\longrightarrow$ Cdc42-act	Act. [46] by Osmostress
Cdc42-act $\longrightarrow$ Cdc42	Deact.
Cdc42 + Ste20 $\longrightarrow$ Cdc42-Ste20	Ass. Req. Cdc42-act [46]
Cdc42-Ste20 $\longrightarrow$ Cdc42 + Ste20	Diss.
$\text{Ste20}^U \longrightarrow \text{Ste20}^P$	Phos. [46] by Cdc42-act
$\text{Ste20}^P \longrightarrow \text{Ste20}^U$	Dephos.
Cdc42 + Ste11 $\longrightarrow$ Cdc42-Ste11	Ass. Req. active Cdc42 [46]
Cdc42-Ste11 $\longrightarrow$ Cdc42 + Ste11	Diss.[1]
$\text{Ste11}^U \longrightarrow \text{Ste11}^P$	Phos. Req. $\text{Ste11-Cdc42-Ste20}^P$ [46]
$\text{Ste11}^P \longrightarrow \text{Ste11}^U$	Dephos.
Sho1 $\longrightarrow$ Sho1-act	Act. [46] by Osmostress
Sho1-act $\longrightarrow$ Sho1	Deact.
Sho1 + Ste11 $\longrightarrow$ Sho1-Ste11	Ass. Req. Sho-act [37, 46]
Sho1-Ste11 $\longrightarrow$ Sho1 + Ste11	Diss.
$\text{Sln1} \longrightarrow \text{Sln1}^P$	Phos. Inh. by Osmostress [27]
Sln1 + Ypd1 $\longrightarrow$ Sln1-Ypd1	Ass. Req. $\text{Sln1}^U,\text{Ypd1}^P$ or $\text{Sln1}^P,\text{Ypd1}^U$ [2] [38]
Sln1-Ypd1 $\longrightarrow$ Sln1 + Ypd1	Diss.
$\text{Sln1}^P\text{-Ypd1}^U \longleftrightarrow \text{Sln1}^U\text{-Ypd1}^P$	Phosphotransfer[3]
Ypd1 + Ssk1 $\longrightarrow$ Ypd1-Ssk1	Ass. Req. $\text{Ypd1}^P,\text{Ssk1}^U$ or $\text{Ypd1}^U,\text{Ssk1}^P$ [2] [38]
Ypd1-Ssk1 $\longrightarrow$ Ypd1 + Ssk1	Diss.
$\text{Ypd1}^P\text{-Ssk1}^U \longleftrightarrow \text{Ypd1}^U\text{-Ssk1}^P$	Phosphotransfer[3]
$\text{Ssk1}^P \longrightarrow \text{Ssk1}^U$	Depho.[4]
Ssk1 + Ssk1 $\longleftrightarrow$ Ssk1-Ssk1	Dimerization [19]
Ssk1 + Ssk2 $\longleftrightarrow$ Ssk1-Ssk2	Ass., Diss.[5]
$\text{Ssk1}^U\text{-Ssk1}^U\text{-Ssk2}^U \longrightarrow$	Phos.
$\longrightarrow \text{Ssk1}^U\text{-Ssk1}^U\text{-Ssk2}^P$	
$\text{Ssk2}^P \longrightarrow \text{Ssk2}^U$	Dephos. Req. unbound Ssk2
Ssk2 + Pbs2 $\longleftrightarrow$ Ssk2-Pbs2	Ass., Diss. [45]
Sho1 + Pbs2 $\longrightarrow$ Sho1-Pbs2	Ass. Req. Sho-act [37, 46]
Sho1-Pbs2 $\longrightarrow$ Sho1 + Pbs2	Diss.
$\text{Pbs2}^{UU} \longrightarrow \text{Pbs2}^{PP}$	Phos. Req. $\text{Ste11}^P\text{-Sho1-Pbs2}$ [26, 46]
	or $\text{Ssk2}^P\text{-Pbs2}$ [45]
$\text{Pbs2}^{PP} \longrightarrow \text{Pbs2}^{UU}$	Dephos.
$\text{Pbs2}^{PP} + \text{Hog1} \longrightarrow \text{Pbs2}^{PP}\text{-Hog1}$	Ass. Req. $\text{Ssk2}^P\text{-Pbs2}^{PP}$ [45]
	or $\text{Ste11}^P\text{-Sho1-Pbs2}^{PP}$ [37]
Pbs2-Hog1 $\longrightarrow$ Pbs2 + Hog1	Diss.
$\text{Pbs2}^{PP}\text{-Hog1}^{UU} \longrightarrow$	Pho. Req. $\text{Ssk2}^P\text{-Pbs2}^{PP}$ [33, 45]
$\longrightarrow \text{Pbs2}^{PP}\text{-Hog1}^{PP}$	Req. $\text{Ste11}^P\text{-Sho1-Pbs2}^{PP}$ [37]
$\text{Hog1}^{PP} \longrightarrow \text{Hog1}^{UU}$	Dephos.[6] [33]
$\text{Hog1}^{PP} + \text{Sho1-act} \longrightarrow$	Ass. [15]
$\longrightarrow \text{Hog1}^{PP}\text{-Sho1-act}$	
Hog1-Sho1 $\longrightarrow$ Hog1 + Sho1	Diss.
$\text{Hog1}^{PP}\text{-Sho1-act} \longrightarrow \text{Hog1}^{PP}\text{-Sho1}$	Inactivation
$\text{Hog1}^{PP}, \text{Osmostress} \longrightarrow \text{Hog1}^{PP}$	Simplified Adaptation

1. Rate can be assumed higher when Ste11 is phoshorylated.

2. Requires reaction partners not bound to members of Sln1-Ypd1-Ssk1-Ssk2 phosphorelay [19].

3. Rate left to right assumed 4 orders of magnitude higher than right to left.

4. Requires Ssk1 to be not bound to Ypd1 and Ssk2.

5. Dissociation rate assumed increased after phosphorylation of Ssk2.

6. Requires Hog1 not bound to Pbs2 [33].

level of abstraction as the reactions in our generic model in Table 1, and each rule may represent several chemical reactions. For detailed information about the language we refer to [9]. A very similar modelling framework is BioNetGen [2]

Cellucidate can accept text input directly from the keyboard, or from a file.

There is also an intuitive graphical user interface (GUI) which can easily be used by biologists to input and edit models. No prior knowledge or experience in mathematical modelling is required for setting up a formal unambiguous description of a system.

We implemented the HOG model in Kappa by translating each generic reaction in Table 1 to a corresponding rule. To illustrate the Kappa syntax and to give a flavour of our model, we give some examples here.

To denote that Sho1 has a state called x that is activated (a) and, furthermore, that Sho1 has a docking site for Ste11, we write

```
Sho1(x~a,Ste11)
```

Similarly, to denote that Ste11 is phosphorylated (p) and has docking sites for both Sho1 and Cdc42, we write

```
Ste11(x~p,Sho1,Cdc42)
```

To model the association of active Sho1 to Ste11, it is reasonable to require that Ste11 is not bound to Cdc42, and that the phosphorylation state of Ste11 is arbitrary. We write the rule as

```
Sho1(x~a,Ste11),Ste11(Sho1,Cdc42) ->
            Sho1(x~a,Ste11!1),Ste11(Sho1!1,Cdc42) @ 2.0
```

Docking is indicated by an index after the reserved symbol !, and the kinetic rate of the rule is defined at the end, after the reserved symbol @. Note that the phosphorylaton state of Ste11 is simply omitted, and that the binding site for Cdc42 on Ste11 is unoccupied on both sides of the arrow.

The formulation is very compact, since only the known interactions and constraints are incorporated. For instance, let us reconsider the example discussed in Section 2. The phosphorylation of Pbs2 by Ste11 is assumed to require that both Ste11 and Pbs2 are bound to Sho1, and furthermore, that Ste11 is phosphorylated. This is encoded as

```
Ste11(x~p,Sho1!1),Sho1(Ste11!1,Pbs2!2),Pbs2(x~u,y~u,Sho1!2) ->
     Ste11(x~p,Sho1!1),Sho1(Ste11!1,Pbs2!2),Pbs2(x~p,y~p,Sho1!2) @ 1.0
```

where x and y refer to a phosphorylation site, u and p refer to unphosphorylated and phosphorylated, respectively. The rule says nothing about the activation status of Sho1 and potential bindings to the other two docking sites of Pbs2 (sites for Hog1 and Ssk2/22). It represents only our current understanding of the system; we do not need to write out all possibilities. When new experimental evidence appears, the model can easily be modified.

An interesting feature of the Kappa system is the possibility to *refine* a certain rule, that is to give an alternative interpretation given some additional requirement [10]. For example, the rate of dissociation of Ste11 from Cdc42 can be assumed higher when Ste11 is phosphorylated. Naturally, each refinement leads to an additional rule with a corresponding kinetic rate, and hence increased complexity of the model.

Implementing models of this kind is error prone, and to debug the model we used the following means:

- The static properties of a model were verified by automatically generated dependency graphs where all agents are nodes and their potential binding partners are edges, a so-called contact map.
- The probability and timing of reaching certain events (the 'firing' of a specific set of rules in the model) given certain initial conditions was verified by simulation of so-called stories.

In addition, we used simulation to judge the qualitative behaviour of the model. Cellucidate uses stochastic simulation and does not rely on internal enumeration of all states. In this way, simulation scales well with the number of rules and possible agents. Especially for very large systems, this becomes prominent [8]. As already mentioned, the parameters of our model are not observable and simulation is not our main objective at this stage. However, in the presented model, we have manually tuned the kinetic rates in order to qualitatively resemble real data. Relative protein abundance data was collected from [14]. The model can be simulated resulting in time-series of number of molecules of various species. A sample plot is given in Fig. 3.

One difficulty arising in simulation models with many states is setting up the initial concentrations. Although not every possible state has to be encoded manually for the rules in Kappa, they have to be for initial conditions. The approach we use here is to run simulations in the unstressed state, starting with unbound and inactive molecules. Assuming the system evolves to a steady state (defined by characteristic distributions of the molecule numbers), we collect the results of a number of simulations. The initial conditions for the simulation under stress are then sampled from these collected results. This approach has the advantage that stochasticity is already included in the initial conditions and that this approach is highly automatable. Drawbacks are that considerable computation is needed to compute the initial steady state and that not all systems necessarily evolve to one steady state (e.g. bistable systems).

4. Discussion

This paper has presented in the language Kappa a formal executable model of the HOG pathway in *S. cerevisiae*, the first such that takes complexes into account. The model, like all Kappa programs, represents the HOG system in terms of agents and rules. It has the desired properties listed earlier:

- It represents current knowledge about the HOG system (summarised as a detailed generic model in Table 1 and as a Kappa model in the Supplementary material) compactly and without the ambiguity of traditional pathway diagrams

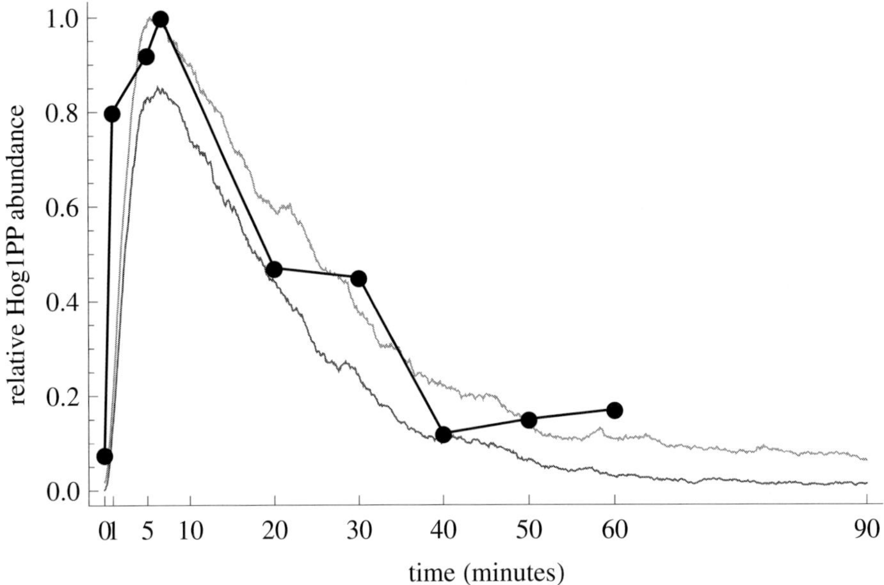

Fig. 3: Simulated and experimental time series for Hog1^{PP} after an osmotic shock. Simulated data was obtained from ten stochastic simulations, and here plotted as $\pm$ one standard deviation. Experimental data (rings, connected by lines) is obtained by the standard Western blot technique.

like in Fig. 2. In particular, our model takes complexes into account without enumerating all states, and captures phosphorylation reactions known to be strictly dependent on the set of species bound to a certain complex.

- The model can easily be edited by biologists, who can use it to anchor discussions about the system. The Kappa formalism allows biological knowledge to be encoded directly into a mathematical model [17]. With the GUI, this encoding requires no prior knowledge of mathematical modelling, and supporting routines for error detection are available. Our model is publicly available for collaborative work on the Cellucidate web site.

- The model includes dynamic information, allowing simulation. Kinetic rates are manually tuned to make simulations resemble experimental data. The model is hence prepared for the natural next step, to be fitted to data. Given current data, this is the best we can do since parameters are not observed when modelling at this level of detail.

4.1. *Constructs Needed to Model Pathways*

Protein interactions alone do not suffice to properly model the dynamics of signalling pathways. Additional model constructs and analysis features may be needed.

- Pathways are naturally integrated with other cellular processes like gene regulation and metabolism, and with the environment. For example, biophysical interactions, gene regulation and glycerol production are key features in adaptation to osmotic shock [22], and cell size is key to cell cycle regulation [1].

- The internal structure of the cell is very complex. Compartments like the cytosol, nucleus and vacuole have individual properties and species concentrations.
- Biological processes at the molecular level are stochastic, and deterministic simulation is not always an adequate approximation. Clearly, hybrid simulation methods are of interest when combining signalling pathways with metabolic systems in one model. When experimental data is gathered from a population of cells (as opposed to single cell experiments), the composition of the population and its temporal development are both stochastic. This population effect is usually not accounted for in mathematical models.

Given this diverse list of requirements, what languages and tools are available and what properties do they have?

4.2. *Comparison of Languages and Softwares*

4.2.1. *Kappa*

Kappa/Cellucidate is very good as far as it goes, but has several shortcomings. It lacks global variables and support for compartments, which can only be implicitly defined, e.g. creating two species $A_{cytosolic}$ and $A_{nuclear}$ each with its own interaction possibilities and transitions from one to the other. Furthermore, the stochastic simulation may be tedious and unnecessary when modelling metabolites with high concentrations.

Finally, Kappa/Cellucidate lacks programmability, so that to program a system with symmetric parts $Q(a)$ and $R(b)$, differing only in the parameter, each part has to be coded separately. In a system like SPiM (below), this can be done in the usual programming way: write a single program $P(x)$, and the two parts are then $P(a)$ and $P(b)$.

When it comes to repeatedly running stochastic simulations, the user has to go through the Cellucidate web-service. Although it is free to use at the moment, Cellucidate and KappaFactory are closed source and extensive use of the web-service might become associated with a fee.

4.2.2. *BioNetGen*

BioNetGen [2, 12] shares many properties with Cellucidate, in particular almost identical input languages. For example, we have relatively easily transcribed the HOG model to the BioNetGen syntax.

The stochastic simulation engine of BioNetGen currently uses a less powerful algorithm than Cellucidate. The new solver under development, nfsim (Sneddon,

Faeder and Emonet, private comm.), implements a generalized version of the algorithm presented in Ref. [47].

In contrast to Cellucidate, BioNetGen is open source, and (fully enumerated) models can be exported to SBML. Models can then be simulated using standard deterministic integration methods. This is an important feature when it comes to parameter estimation, where simulation speed often is crucial.

BioNetGen and its web interface GetBonNie (`http://getbonnie.cs.unm.edu`) feature less sophisticated GUIs compared to KappaFactory/Cellucidate. However, all options of the software are directly available from the command line, making repeated simulations of stochastic models easy to set up.

4.2.3. *SPiM*

The Stochastic Pi Machine (SPiM) [3, 35, 36, 43] runs the stochastic version of the the pi-calculus [31, 32]. The aim of the pi-calculus is to describe concurrent computations whose configuration may change during the computation. That the stochastic pi-calculus could be used for biochemistry was first discovered by Aviv Regev [39, 41].

The SPiM programming language allows parameterisation of symmetric code, as mentioned above. It has global variables, though not fully dynamic ones. Compartments (even changing ones) can be represented, but coded in a way that might not be transparent to a biologist. Similarly, a link between two molecules A and B can be encoded in various ways, by moving both into new states where a parameter records the linkage. We can choose whether to say just that A is linked on this contact point x, or add details such as to what kind of molecule it is linked to, or even that it is specifically linked to B. Thus the programmibility is a gain, but at the cost of direct representation of chemistry.

4.2.4. *Rule-based Modelling in PottersWheel*

PottersWheel [28] (PW) is is based on ODEs and deterministic simulation, and includes a range of analysis tools like parameter estimation, sensitivity analysis and model discrimination. Besides, PW offers a syntax for describing complexes by rule based reactions. The full set of ODEs is then automatically generated and used in simulations.

A fundamental difference between Kappa and rule-based modelling in PW is that Kappa by default considers all possible reactions, while in PW the alternatives must be explicitly defined. In general, the rule-based modelling in PW offers the same flexibility as Kappa, but the code is less compact, and requires additional manual work.

As a modelling software tool, PW offers more general functionality compared to Kappa/Cellucidate, e.g. algebraic equations, arbitrary input functions, and compartments. Besides, PW is very strong on model analysis and parameter estimation, while it is currently lacking support for stochastic simulation and a GUI for rule-based modelling.

4.3. *Concluding Remarks*

Mathematical models enable a formal representation of biological systems and thorough analysis of hypotheses generated from experiments. The success of such models demands further refinement of mathematical methods and computational tools, as well as of experimental methods.

There are now several modelling frameworks, and it is important to choose the one appropriate to the purpose at hand, to the detail at which the system is described, and to the nature of the system. Our Kappa model of the HOG system is valuable as an unambiguous description of the protein interactions involved, and therefore as a basis for discussion. It cannot currently be used to study the complete adaptation of a yeast cell to hyperosmotic shock, from sensing the biophysical changes in turgor changes to accumulation of intracellular glycerol. But this is exactly the strength of such descriptions. Each allows us to 'zoom in' on a specific part of the problem. How to integrate the results in a higher level description of the system is a topic for new research.

Mathematical modelling of signalling pathways requires a large amount of experimental data, computation power, and the coordinated work of many scientists. This coordination will eventually have to go beyond formulation of common standards and storage of models in databases. One way is to work directly on a common, executable model that unambiguously represents current knowledge. Our HOG model in Kappa is such a model, and the web-based approach of Cellucidate allows it to be directly annotated and discussed by multiple users, for example to jointly incorporate new implications as those mentioned in [25].

Acknowledgments

The project was done as part of a Marie Curie EST school in Göteborg 2008. CK was funded by the International Research Training Group *Genomics and Systems Biology of Molecular Networks*, supported by the German Research Foundation (DFG). PG was supported by the Göteborg University quantitative biology platform and the Swedish Strategic Research Foundation through Göteborg Mathematical Modeling Center. Thanks to Vincent Danos for introducing us to Kappa, to James R. Faeder for support on BioNetGen and to Jörg Schaber for comments on the model. *Author contributions:* CK and PG conceived the modelling concept, and constructed the basic model. CK significantly improved the model and tuned it to data. KVSP contributed technical expertise and essential modelling language background information. EK contributed experience with the Hog-Pathway and modeling thereof. PG wrote the paper with significant contributions from CK and KVSP.

References

[1] Barberis, M., Klipp, E., Vanoni, M., Alberghina, L., Cell Size at S Phase Initiation: an Emergent Property of the G1/S Network, *PLoS Comput. Biol.*, 3(4): e64, 2007.

[2] Blinov, M., Faeder, J., Goldstein, B., Hlavacek, W., Bionetgen: software for rule-based modeling of signal transduction based on the interactions of molecular domains, *Bioinformatics*, 20(17):3289–3291, 2004.

[3] Blossey, R., Cardelli, L., Phillips, A., A Compositional Approach to the Stochastic Dynamics of Gene Networks, *Transactions in Computational Systems Biology (TCSB)*, 3939:99–122, 2006.

[4] Borisov, N., Markevich, N., Hoek, J., Kholodenko, B., Signaling through receptors and scaffolds: independent interactions reduce combinatorial complexity, *Biophys. J.*, 89(2):951–966, 2005.

[5] Borisov, N., Markevich, N., Hoek, J., Kholodenko, B., Trading the micro-world of combinatorial complexity for the macro-world of protein interaction domains, *Biosystems*, 83(2-3):152–166, 2006.

[6] Choi, M.Y., Kang, G.Y., Hur, J.Y., Jung, J.W., Kim K.P., Park, S.H., Analysis of Dual Phosphorylation of Hog1 MAP Kinase in Saccharomyces cerevisiae Using Quantitative Mass Spectrometry, *Mol. Cells*, 26:200–205, 2008.

[7] Conzelmann, H., Saez-Rodriguez, J., Sauter, T., Kholodenko, B., Gilles, E., A domain-oriented approach to the reduction of combinatorial complexity in signal transduction networks, *BMC Bioinformatics*, 7:3, 2006.

[8] Danos, V., Feret, J., Fontana, W., Krivine, J., Scalable Simulation of Cellular Signaling Networks, *APLAS 2007, the Fifth ASIAN Symposium on Programming Languages and Systems*, 2007.

[9] Danos, V., Feret, J., Fontana, W., Harmer, R., Krivine, J., Rule-based modelling of cellular signalling, *CONCUR, the 18th International Conference on Concurrency Theory*, 2007.

[10] Danos, V., Feret, J., Fontana, W., Harmer, R., Krivine, J., Rule-based modelling, symmetries, refinements, *Lecture Notes in Computer Science*, 5054:103–122, 2008.

[11] Dihazi, H., Kessler, R., Eschrich, K., High osmolarity glycerol (HOG) pathway-induced phosphorylation and activation of 6-phosphofructo-2-kinase are essential for glycerol accumulation and yeast cell proliferation under hyperosmotic stress, *J. Biol. Chem.*, 279:23961–23968, 2004.

[12] Faeder, J.R., Blinov, M.L., Hlavacek, W.S., Rule-based modeling of biochemical systems with BioNetGen,*Methods Mol. Biol*, 500:113–167, 2009.

[13] Gennemark, P., Nordlander, B., Hohmann, S., Wedelin, D., A simple mathematical model of adaptation to high osmolarity in yeast, *In Silico Biol.*, 6(3):193–214, 2006.

[14] Ghaemmaghami S., Huh W.K., Bower K., Howson R.W., Belle A., Dephoure N., O'Shea E.K., Weissman J.S., Global analysis of protein expression in yeast, *Nature*, 425(6959):737–741, 2003.

[15] Hao, N., Behar, M., Parnell, S.C., Torres, M.P., Borchers, C.H., Elston, T.C., Dohlman, H.G., A systems-biology analysis of feedback inhibition in the Sho1 osmotic-stress-response pathway, *Curr. Biol.*, 17(8):659–667, 2007.

[16] Hersen, P., McClean, M.N., Mahadevan, L., Ramanathan, S. Signal processing by the HOG MAP kinase pathway, *Proc Natl Acad Sci U S A.*, 105(20):7165–7170, 2008.

[17] Hlavacek W.S., How to deal with large models?, *Molecular Systems Biology*, 5(240), 2008.

[18] Hohmann, S., Osmotic stress signaling and osmoadaptation in yeasts, *Microbiol Mol Biol Rev.*, 66(2):300–372, 2002.

[19] Horie, T., Tatebayashi, K., Yamada, R., Saito, H., Phosphorylated Ssk1 prevents

unphosphorylated Ssk1 from activating the Ssk2 mitogen-activated protein kinase kinase kinase in the yeast high-osmolarity glycerol osmoregulatory pathway, *Mol. Cell Biol.*, 28(17):5172–5183, 2008.

[20] Hucka, M., Finney, A., Sauro, H.M., Bolouri, H., Doyle, J.C., *et al.*, The Systems Biology Markup Language (SBML): A medium for representation and exchange of biochemical network models, *Bioinformatics*, 19(4):524–531, 2003.

[21] Hucka, M., Finney, A., Hoops, S., Keating, S., Le Novere, N., Systems Biology Markup Language (SBML) Level 2: Structures and Facilities for Model Definitions. SBML Level 2 Version 3, Release 2, 26 September 2007. Section 8.1. http://sbml.org, 2007.

[22] Klipp, E, Nordlander, B, Krüger, R, Gennemark, P, Hohmann, S., Integrative model of the response of yeast to osmotic shock, *Nat Biotechnol.*, 23(8):975–982, 2005.

[23] Kühn, C., Petelenz, E., Nordlander B., Jörg Schaber, Hohmann, S., Klipp, E., Exploring the Impact of Osmoadaptation on Glycolysis Using Time-Varying Reponse-Coefficients, *Genome Informatics*, 20:77–90, 2008.

[24] Levchenko A., Bruck J., Sternberg P.W., Scaffold proteins may biphasically affect the levels of mitogen-activated protein kinase signaling and reduce its threshold properties, *PNAS*, 97(11):5818–5823, 2000.

[25] Macia, J., Regot, S., Peeters, T., Conde, N., Solé, R., Posas, F., Dynamic Signaling in the Hog1 MAPK Pathway Relies on High Basal Signaling Transduction, *Science Signaling*, 2(63), ra13, 2009.

[26] Maeda, T., Takekawa, M., Saito, H., Activation of yeast PBS2 MAPKK by MAPKKKs or by binding of an SH3-containing osmosensor, *Science*, 269(5223):554–558, 1995.

[27] Maeda, T., Wurgler-Murphy, S.M., Saito, H., A two-component system that regulates an osmosensing MAP kinase cascade in yeast, *Nature*, 369:242–245, 1994.

[28] Maiwald, T., Timmer, J., Dynamical modeling and multi-experiment fitting with PottersWheel, *Bioinformatics*, 24(18):2037–2043, 2008.

[29] Mettetal, J.T., Muzzey, D., Gomez-Uribe, C., van Oudenaarden, A., The frequency dependence of osmo-adaptation in Saccharomyces cerevisiae, *Science*, 319(5862):482–484, 2008.

[30] Milner, R., *A Calculus of Communicating Systems*, Springer Verlag, ISBN 0-387-10235-3, 1980.

[31] Milner, R., Parrow, J., Walker, D., A calculus of mobile processes, *Information and Computation*, 100(100), 1–40, 1992.

[32] Milner, R., *Communicating and Mobile Systems: the Pi-Calculus*, Cambridge Univ. Press, ISBN 0-521-65869-1, 1999.

[33] Murakami, Y., Tatebayashi, K., Saito, H., Two adjacent docking sites in the yeast Hog1 mitogen-activated protein (MAP) kinase differentially interact with the Pbs2 MAP kinase kinase and the Ptp2 protein tyrosine phosphatase, *Mol. Cell. Biol.*, 28(7):2481–2494, 2008.

[34] Netz, D.J., Pierik, A.J., Stümpfig, M., Mühlenhoff, U., Lill, R., The Cfd1-Nbp35 complex acts as a scaffold for iron-sulfur protein assembly in the yeast cytosol, *Nat. Chem. Biol.*, 3(5):278–286, 2007.

[35] Phillips, A., Cardelli, L., Castagna, G., A Graphical Representation for Biological Processes in the Stochastic Pi-calculus, *Transactions in Computational Systems Biology (TCSB)*, 4230:123–152, 2006.

[36] Phillips, A., Cardelli, L., Efficient, Correct Simulation of Biological Processes in the Stochastic Pi-calculus, In Proceedings of Computational Methods in Systems Biology (CMSB'07), 4695, 184–199, 2007.

[37] Posas, F., Saito, H., Osmotic activation of the HOG MAPK pathway via Ste11p

MAPKKK: scaffold role of Pbs2p MAPKK, *Science*, 276(5319):1702–1705, 1997.

[38] Posas, F., Wurgler-Murphy, S.M., Maeda, T., Witten, E.A., Thai, T.C., Saito, H., Yeast HOG1 MAP kinase cascade is regulated by a multistep phosphorelay mechanism in the SLN1-YPD1-SSK1 "two-component" osmosensor, *Cell*, 86(6):865–875, 1996.

[39] Priami, C., Regev, A., Shapiro, E.Y., Silverman, W., Application of a stochastic name-passing calculus to representation and simulation of molecular processes, *Inf. Process. Lett.*, 80(1):25-31, 2001.

[40] Reiser, V., Raitt, D. C. and Saito, H., Yeast osmosensor Sln1 and plant cytokinin receptor Cre1 respond to changes in turgor pressure, *J. Cell Biol.*, 161:1035-1040, 2003.

[41] Regev, A., Silverman, W., Shapiro, E.Y., Representation and Simulation of Biochemical Processes Using the pi-Calculus Process Algebra, *Pacific Symposium on Biocomputing*, 6:459–470, 2001.

[42] Rischitor, P.E., May, K.M., Hardwick, K.G., Bub1 is a fission yeast kinetochore scaffold protein, and is sufficient to recruit other spindle checkpoint proteins to ectopic sites on chromosomes, *PLoS ONE*, 2(12): e1342, 2007.

[43] SPiM: http://research.microsoft.com/~aphillip/spim

[44] Tamas, M.J., Rep, M., Thevelein, J.M., Hohmann, S., Stimulation of the yeast high osmolarity glycerol (HOG) pathway: evidence for a signal generated by a change in turgor rather than by water stress, *FEBS Lett.*, 472(1):159-65, 2000.

[45] Tatebayashi, K., Takekawa, M., Saito H., A docking site determining specificity of Pbs2 MAPKK for Ssk2/Ssk22 MAPKKKs in the yeast HOG pathway, *The EMBO Journal*, 22:3624-3634, 2003.

[46] Tatebayashi, K., Yamamoto, K., Tanaka, K., Tomida, T., Maruoka, T., Kasukawa, E., Saito, H., Adaptor functions of Cdc42, Ste50, and Sho1 in the yeast osmoregulatory HOG MAPK pathway, *EMBO J.*, 25(13):3033-44, 2006.

[47] Yang, J., Monine, M.I., Faeder, J.R., Hlavacek, W.S., Kinetic Monte Carlo method for rule-based modeling of biochemical networks , *Phys. Rev. E.*, 78:031910, 2008.

PREDICTION OF REGULATORY TRANSCRIPTION FACTORS IN T HELPER CELL DIFFERENTIATION AND MAINTENANCE

YUE-HIEN LEE[1,2]

yuehien@gmail.com

MANUELA BENARY[1]

manuela.benary@cms.hu-berlin.de

RIA BAUMGRASS[2]

baumgrass@drfz.de

HANSPETER HERZEL[1]

h.herzel@biologie.hu-berlin.de

[1]*Institute for Theoretical Biology, Humboldt University of Berlin, Invalidenstr. 43, 10115 Berlin, Germany*
[2]*German Rheumatism Research Centre, Chariteplatz 1, 10117 Berlin, Germany*

Naive T-helper (Th) cells differentiate into distinct lineages including Th1, Th2, Th17 and regulatory T (Treg) cells. Each of these Th-lineages has specific functions in immune defense and T cell homeostasis. Th cell fate decisions and commitment are dependent on the kind and strength of T cell stimulation and the subsequent gene expression profiles.

Our analysis targeted the identification of new regulatory transcription factor binding sites (TFBSs) in the promoter regions of up- and down-regulated genes in Treg cell differentiation and lineage maintenance. For this approach we compared different gene groups from global gene expression studies with background models of randomly selected genes to identify significantly overrepresented TFBSs.

Results of our analysis suggest that Ets and IRF family members contribute to the regulation of the initial induction of Treg cells. Furthermore, we identified the overrepresented TFBS-pairs Runx-NFAT and GATA3-Foxp3 in Treg specific genes and Foxp3 dependent genes, respectively. Interestingly, previous studies have observed functional interactions of both TFBS-pairs in T cells. This study provides a starting point for further investigations to elucidate the transcriptional network in Treg cells.

Keywords: regulatory T cells; differentiation; promoter analysis; transcription factors.

1. Introduction

T-helper (Th) cells are the central players of adaptive immune responses in mammalians. There are different effector subsets of Th cells, such as Th1, Th2, and Th17 cells, as well as Treg cells. Each of these Th cell subsets has specific functions in pathogen defense and regulation of immune reaction. Dysregulation of T cell activation and differentiation can lead to autoimmune diseases and allergies. Therefore, it is important to understand how these processes are regulated. Transcription factors (TFs) and cytokines are crucial orchestrators of Th cell fate decision processes. Often, not only one master transcription factor, but a transcription factor network is essential for induction and maintenance of different Th-subpopulations.

Treg cells represent a unique population of Th cells that actively inhibit other immune cells by different mechanisms. The lineage specifying transcription factor Foxp3 is a hallmark of the Treg cell lineage [13]. Ectopic overexpression of Foxp3 in resting Th cells successfully converts these into functional Treg cells [9]. Mutations or deletion of Foxp3 cause severe autoimmune disorders [6]. Peripheral naive Th cells can differentiate into Foxp3$^+$ Treg cells by simultaneous stimulation of the T cell receptor and the TGF-β receptor in vivo and ex vivo [13, 39]. In addition, TGF-β is necessary for the maintenance and in vivo expansion of peripheral Treg cells [15, 21]. However, the molecular mechanisms of TGF-β-dependent Treg cell generation and maintenance are largely unknown.

Few transcription factors have been reported to promote TGF-β-dependent de novo induction of Foxp3 expression. Among them are TIEG1 (also named Kfl10) [36], NFAT and SMAD3 [35]. Other transcription factors are supposed to act as negative regulators of Foxp3 expression, such as GATA3 [20] and STAT6 [33].

To discover more of such regulators of Foxp3 expression as well as Treg cell induction and maintenance, we analysed different gene groups for overrepresented binding sites of transcription factors and pairs of transcription factors.

2. Materials and Methods

Recent studies showed that many transcription factors bind symmetrically close to the TSS [3, 32]. Therefore, we decided to select the region -500 to +500 bp around the transcription start site (TSS) for each gene and call this region hereafter promoter region. Promoter regions were extracted from the Ensembl database [11]. All 610 vertebrate binding site motifs represented by positional weight matrices (PWMs) were taken from TRANSFAC 12.1 [24]. For prediction of single binding sites the TFBS perl module from Lenhard et al. [19] was used.

Selection of Gene Groups

To identify new regulating transcription factors in Treg cells we selected four gene groups (Table 1) from global gene expression studies and analyzed transcription factor binding sites in the genes within the depicted groups. Genes of the first two groups were extracted from our own global gene expression studies. The first gene group contains genes differentially expressed in naive T-helper cells after stimulation with anti-CD3/CD28 antibodies for three hours (Table 1, gene group STIM). This group comprises immediate early genes of Th cell activation. The second group constitutes genes which are differentially expressed after treatment of cells with low-dose TGF-β (0.2 ng/ml) and low-dose Cyclosporine A (CsA, 5 nM). These conditions promote the induction of Foxp3$^+$ regulatory Th cells. The third group contains Foxp3 dependent genes and the fourth group Treg specific genes. Both groups were extracted from four global gene expression and chromatin-immuno-precipitation (ChIP) studies [12, 25, 31, 40].

Table 1: Selected gene groups with number of genes in the group (# Genes) and the average GC-content of the promoter regions inside the group (GC %).

Gene group	Description	GC %	# Genes	Targets of our TFBS search
STIM	genes in Th cell activation	58.38	2871	regulatory TFs in stimulated naive Th cells
TGF-β /CsA	genes in Treg cell induction	56.86	101	regulatory TFs during initial Treg cell induction
TREG	Treg specific genes	57.12	543	regulatory TF pairs for Treg cell maintenance
FOXP3	Foxp3 dependent genes	57.6	601	regulatory TF pairs in Foxp3 dependent genes

Selection of Transcription Factors

To search for combinatorial control of TFs in Treg cells, we selected TFs with known regulatory functions in either T cell development or T cell function (Table 2).

Table 2: Selected transcription factors for combinatorial analyses.

Transcription factors	Function
AHR	activates differentiation of Treg and Th17 cells [26]
AP-1, NFAT, NF-κB	major TFs activated by T cell stimulation [30]
C/EBP	mediator of Il-17 signalling [29]
CREB, ATF	CREB/ATF binding sites are important for T cell receptor induced expression of Foxp3 [17]
EGR	EGR-2 and EGR-3 have negative effects on T cell activation [28]
Foxp3	master transcription factor in Treg cells [13]
GATA3	master transcription factor in Th2 cells [41]
p300, SMAD3, TGIF	components of TGF-β signalling pathway [23]
RAR-α/RXR-α	prevent the STAT6 inhibition of Foxp3 [33]
RORα	regulates the differentiation of Th17 cells [8]
RORγt	master transcription factor in Th17 cells [8]
Runx1	inhibits with Foxp3 and promotes with RORγt Th17 cell differentiation [38]
STAT3	main TF of Il-6 signalling pathway [8]
STAT5	regulates the Foxp3 expression directly [37]
TIEG	activates TGF-β-induced Foxp3 expression [36]

Determination of an Overrepresented Motif in a Gene Group

First, we generated an empirical distribution of predicted number of binding sites for the motif using randomly selected genes. Next, with this empirical distribution we calculated for each gene in the group the p-value to observe the predicted number of binding sites for the motif. P-values below 0.05 are considered as significant. Based on these p-values we applied two commonly used statistical hypothesis tests, Fisher's exact test and Z-test, to determine an overrepesentation of the motif in the group. For the Z-test the z-sores were calculated by comparing the sum of p-values of genes to a background model. We chose to use the sum of p-values in order to avoid single genes having to much influence. According to this approach, motifs with z-scores smaller than -3 are regarded as overrepresented in the group. We chose for Fisher's exact test the same significance level, corresponding to 0.0027.

Compensating GC-Rich Promoter Regions

Promoter regions of the selected gene groups are GC-rich (Figure 1). GC-rich promoters have also been observed in previous studies, e.g. by Bozek et al. [5] in clock-controlled genes. Potentially, GC-rich promoters have the ability to ease gene transcription, since CpG-island promoters are facilitating constitutive gene expression or rapid gene induction without requiring SWI/SNF nucleosome remodeling complexes [27].

If we quantify TFBS overrepresentation with our approach or with F-MATCH [16] using arbitrary mouse genes as a background model, we find many GC-rich motifs. This can be regarded as an artefact due to the high GC-content of our gene groups. Consequently, in this paper we apply GC-matched background models as in Bozek et al. [5].

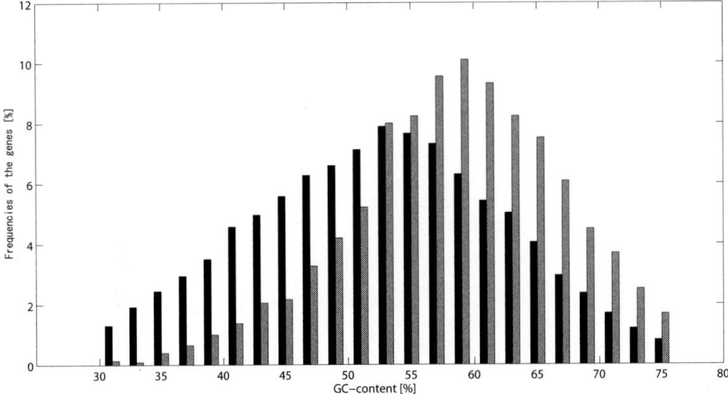

Fig. 1: GC-content distribution of analysed promoter regions (light) and promoter regions from the background model (black).

3. Results

Overrepresented Motifs in the Gene Group of Activated Naive Th Cells

Overrepresented motifs were determined by Fisher's exact test and by the Z-test. By comparing results of Fisher's exact test to the results of the Z-test, we observed that all predictions by Fisher's exact test were obtained also by the Z-test. The lists contain overrepresented motifs with low and high GC-content. This indicates that we compensated the high GC-content successfully with our background model. We present in Table 3 overrepresented motifs detected by both tests. The transcription factors associated to the motifs point to links between T cell activation and cAMP and ERK signalling, cell cycle and differentiation. Interestingly, many of these transcription factors are also predicted to regulate clock-controlled genes [5].

Table 3: Overrepresented motifs in group STIM (p-values from Fisher's exact test). The comments are based on TRANSFAC and Gene Ontology [2].

Motif	TF	Comment	P-value
V$E2F_Q2	E2F family	cell cycle regulation	3.6271e-06
V$ELK1_02	Elk-1	ERK/MAPK target	7.9353e-06
V$NFY_01	NF-Y	regulates MHC II genes	2.1575e-05
V$NRF2_01	Nrf-2	stress response	2.5528e-05
V$CETS1P54_03	c-Ets-1	ERK/MAPK target	3.9109e-05
V$CREBATF_Q6	CREB, ATF	cAMP signalling	0.0001
V$NRF1_Q6	Nrf-1	induces MEF-2a	0.0002
V$HMGIY_Q3	HMG-I(Y)	lymphocyte differentiation	0.0003
V$ATF_01	ATF	cAMP signalling	0.0006
V$CDX_Q5	Cdx	cell differentiation	0.0013
V$CIZ_01	CIZ	myeloid cell differentiation	0.0016
V$MMEF2_Q6	MEF-2a	muscle specific	0.0016
V$CREB_02	CREB	cAMP signalling	0.0018
V$WHN_B	Foxn1	T cell proliferation	0.0019
V$AP2GAMMA_01	AP-2γ	retinoic acid target	0.0019
V$AHR_Q5	AHR	differentiation of T cells	0.0021
V$IRF_Q6_01	IRF family	differentiation of T cells	0.0022

Transcription Factors of the Ets Family and IRF Family are Overrepresented in the Gene Group of Treg Cell Induction

We identified four motifs (IRF8, IRF family, PU.1 and Elf1) as overrepresented in the gene group TGF-β/CsA (see Figure 2). The binding site motif of IRF8 is quite similar to the one of IRF family. Note, that reverse complementary motifs represent the same binding site on the opposite DNA strand. The two similar motifs of PU.1 and Elf1 belong to the Ets family. In fact, it has been observed that PU.1 can bind to Elf1 binding sites [4]. PU.1 and Elf1 seem to differ from other Ets family members in their DNA binding specificity [4].

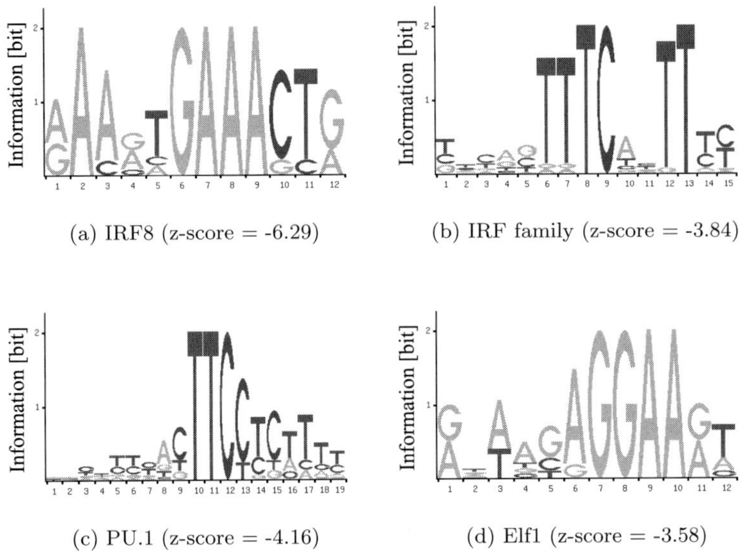

(a) IRF8 (z-score = -6.29)

(b) IRF family (z-score = -3.84)

(c) PU.1 (z-score = -4.16)

(d) Elf1 (z-score = -3.58)

Fig. 2: Overrepresented motifs in the gene group TGF-β/CsA.

Combinatorial Analysis Revealed GATA3-Foxp3 and Runx-NFAT as Potential Regulatory Complexes

We used the transcription factors listed in Table 2 and the transcription factors found in the previous section. Altogether we had 25 motifs, 20 motifs from TRANSFAC, and 5 motifs based on sites from the literature such as Marson et al. [22]. Out of the 25 single motifs we created 325 motif pairs by combination. The distance between single binding sites in a motif pair was 1 to 15 bp in the initial screen.

The combinatorial analysis consisted of three steps. In the first step we searched in the gene groups TREG and FOXP3 for overrepresented motif pairs in the same way as for the single motifs. We discovered 8 overrepresented motif pairs in the group TREG and 18 in the group FOXP3. In the next step the overrepresented motif pairs where filtered using a heuristic criterion to reduce the number of predictions. Only motif pairs with at least one motif having a z-score greater than zero are chosen for further analysis. Thus, we focused on motif pairs where at least one motif was not overrepresented. After this step we had 4 overrepresented motif pairs left. IRF8-STAT3, NFAT-STAT3 and GATA3-Foxp3 in the group FOXP3 and Runx1-NFAT in the group TREG. Since Runx1, Runx2 and Runx3 have almost identical DNA-binding domains [1], we refer from now on to Runx-NFAT. In the last step we conducted a distance analysis for the 4 remaining motif pairs (Figure 3). We assume that motif pairs with preferred short distances are particularly likely to interact. We divided the range of 0 to 200 bp into 20 bp intervals and repeated for each interval the overrepresentation analysis. As a result of the distance analysis only Runx-NFAT and GATA3-Foxp3 have a clear preference for short distances. Thus, our combinatorial analysis suggests that GATA3 and Foxp3 as well as Runx and NFAT interact in short distances with each other.

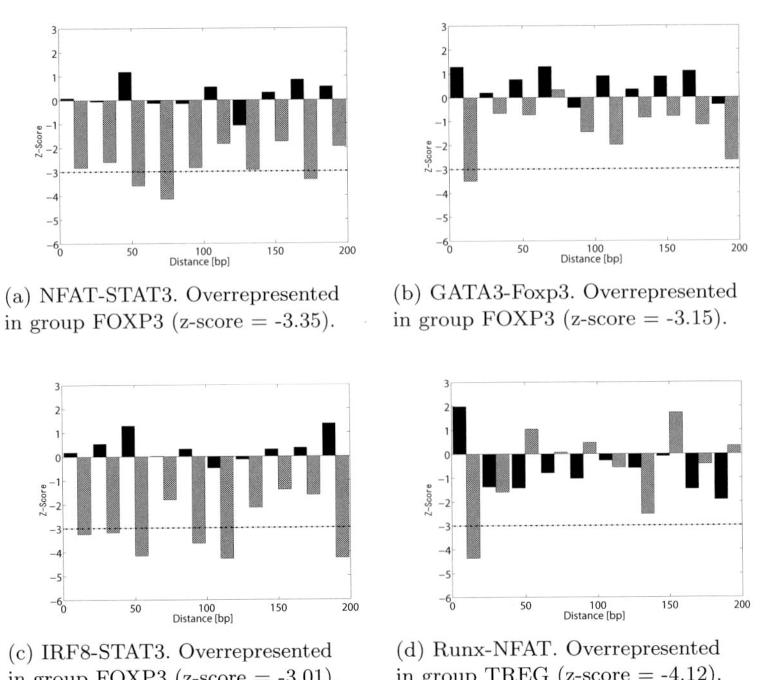

(a) NFAT-STAT3. Overrepresented in group FOXP3 (z-score = -3.35).

(b) GATA3-Foxp3. Overrepresented in group FOXP3 (z-score = -3.15).

(c) IRF8-STAT3. Overrepresented in group FOXP3 (z-score = -3.01).

(d) Runx-NFAT. Overrepresented in group TREG (z-score = -4.12).

Fig. 3: Distance analysis was applied for the overrepresented motif pairs in 20 bp intervals (red = target gene group, dark = control group). For NFAT-STAT3 and IRF8-STAT3 no distances were preferred over others, whereas Runx-NFAT and GATA3-Foxp3 have clear preferences for short distances.

4. Discussion

The transcriptional network regulating differentiation, function, and maintenance of Treg cells is largely unknown. In this study we used a bioinformatic approach to identify transcription factors involved in the regulation of these processes in Treg cells. For this purpose we analysed transcription factor binding sites in promoter regions to identify significantly overrepresented motifs in comparison to a background model. Importantly, in accordance with Bozek et al. [5], we compensated the high GC-content of the analysed promoter regions by using a GC-matched background model to reduce false positive results.

The analysis of immediate early genes in Treg cell induction suggests that transcription factors of the Ets family, PU.1 and Elf1, and of the IRF family are involved in Treg cell differentiation processes. Experimental data for IRF4, one IRF family member, support our results. It is already known that IRF4 is essential for the differentiation of Th17 cells [8]. These data do not rule out specific functions of other IRF family members in Treg cells. The role of Ets family transcription factors in peripheral Th cell differentiation is largely unknown. However, it has been reported that Elf1 regulates gene expression during T cell activation [34] and that transiently expressed PU.1 might inhibit Foxp3 during initial Th2 differentiation [10]. Interestingly, cell fate decisions of T cell progenitors in the thymus are controlled by PU.1 [14]. Therefore, an important role of PU.1 in peripheral Treg cell linage decision is possible, too.

Furthermore, we suppose from our results that the transcription factor pairs GATA3-Foxp3 and Runx-NFAT are important for Treg cells. Indeed, experimental data of Dardalhon et al. [7] showed already in Th2 cells that GATA3-Foxp3 interaction inhibited GATA3 mediated transactivation of Il-5. They suggested an inhibition of other GATA3 targets as well. Interestingly, there is already a hint that GATA3 and Foxp3 are transiently coexpressed during the second day of Treg cell induction [20]. Therefore, it would be interesting to investigate whether GATA3-Foxp3 interaction promotes Treg cell lineage decision by inhibition of Th2 cell development. This could be one mechanism how the master transcription factors help to shape Th cell lineage decisions.

Not only GATA3-Foxp3 but also Runx-NFAT was already described as an interaction pair in Th2 cells inhibiting IL-4 expression [18]. Our results suggest that transcription factors of the Runx family interact with NFAT to regulate Treg cell specific genes. Therefore, this interaction should be analysed in more detail in the context of Treg cell induction and maintenance.

Our bioinformatics analyses of global gene expression data revealed several new promising starting points for further investigations to elucidate the transcriptional networks regulating Treg cell differentiation and maintenance.

Acknowledgments

The work was supported by the IRTG and by a grant of the German Ministry of Education and Research within the Forsys Partner Network to H.H. and R.B..

References

[1] Alarcon-Riquelme, M.E., A RUNX trio with a taste for autoimmunity, *Nature Genetics*, 35:299–300, 2003.

[2] Ashburner, M., et al., Gene ontology: tool for the unification of biology. The Gene Ontology Consortium, *Nature Genetics*, 25:25–29, 2000.

[3] Birney, E., et al., Identification and analysis of functional elements in 1% of the human genome by the ENCODE pilot project, *Nature*, 447:799–816, 2007.

[4] Bockamp, E.O., et al., Transcriptional regulation of the stem cell leukemia gene by PU.1 and Elf-1, *J. Biol. Chem.*, 273:29032–29042, 1998.

[5] Bozek, K., et al., Regulation of clock-controlled genes in mammals, *PloS One*, 4:e4882, 2009.

[6] Brunkow, M.E., et al., Disruption of a new forkhead/winged-helix protein, scurfin, results in the fatal lymphoproliferative disorder of the scurfy mouse, *Nature Genetics*, 27:68–73, 2001.

[7] Dardalhon, V., et al., IL-4 inhibits TGF-beta-induced Foxp3+ T cells and, together with TGF-beta, generates IL-9+ IL-10+ Foxp3(-) effector T cells, *Nature Immunology*, 9:1347–1355, 2008.

[8] Dong, C., Th17 cells in development: an updated view of their molecular identity and genetic programming, *Nature Reviews Immunology*, 8:337–348, 2008.

[9] Fontenot, J.D., Gavin, M.A., Rudensky, A.Y., Foxp3 programs the development and function of CD4+CD25+ regulatory T cells, *Nature Immunology*, 4:330–336, 2003.

[10] Hadjur, S., et al., IL4 blockade of inducible regulatory T cell differentiation: the role of Th2 cells, Gata3 and PU.1, *Immunology Letters*, 122:37–43, 2009.

[11] Hammond, M.P., Birney, E., Genome information resources - developments at Ensembl, *Trends in Genetics: TIG*, 20:268–272, 2004.

[12] Hill, J.A., et al., Foxp3 transcription-factor-dependent and -independent regulation of the regulatory T cell transcriptional signature, *Immunity*, 27:786–800, 2007.

[13] Hori, S., Nomura, T., Sakaguchi, S., Control of regulatory T cell development by the transcription factor Foxp3, *Science*, 299:1057–1061, 2003.

[14] Huang, G., et al., PU.1 is a major downstream target of AML1 (RUNX1) in adult mouse hematopoiesis, *Nature Genetics*, 40:51–60, 2008.

[15] Huber, S., et al., Cutting edge: TGF-beta signaling is required for the in vivo expansion and immunosuppressive capacity of regulatory CD4+CD25+ T cells, *Immunology*, 173:6526–6531, 2004.

[16] Kel, A., et al., Beyond microarrays: Finding key transcription factors controlling signal transduction pathways, *BMC Bioinformatics*, 7 Suppl 2:S13, 2006.

[17] Kim, H., Leonard, W.J., CREB/ATF-dependent T cell receptor-induced FoxP3 gene expression: a role for DNA methylation, *Medicine*, 204:1543–1551, 2007.

[18] Lee, S.H., et al., Runx3 inhibits IL-4 production in T cells via physical interaction with NFAT, *Biochem. and Biophys. Research Communications*, 381:214–217, 2009.

[19] Lenhard, B., Wasserman, W.W., TFBS: computational framework for transcription factor binding site analysis, *Bioinformatics*, 18:1135–1136, 2002.

[20] Mantel, P., et al., GATA3-driven Th2 responses inhibit TGF-beta1-induced FOXP3 expression and the formation of regulatory T cells, *PLoS Biology*, 5:e329, 2007.

[21] Marie, J.C., et al., TGF-beta1 maintains suppressor function and Foxp3 expression in CD4+CD25+ regulatory T cells, *Medicine*, 201:1061–1067, 2005.

[22] Marson, A., et al., Foxp3 occupancy and regulation of key target genes during T-cell stimulation, *Nature*, 445:931–935, 2007.

[23] Massagu, J., Chen, Y.G., Controlling TGF-beta signaling, *Genes & Development*, 14:627–644, 2000.

[24] Matys, V., et al., TRANSFAC and its module TRANSCompel: transcriptional gene regulation in eukaryotes, *Nucleic Acids Research*, 34:D108–110, 2006.

[25] McHugh, R.S., et al., CD4(+)CD25(+) immunoregulatory T cells: gene expression analysis reveals a functional role for the glucocorticoid-induced TNF receptor, *Immunity*, 16:311–323, 2002.

[26] Quintana, F.J., et al., Control of Treg and Th17 cell differentiation by the aryl hydrocarbon receptor, *Nature*, 453:65–71, 2008.

[27] Ramirez-Carrozzi, V.R., et al., A unifying model for the selective regulation of inducible transcription by CpG islands and nucleosome remodeling, *Cell*, 138:114–128, 2009.

[28] Safford, M., et al., Egr-2 and Egr-3 are negative regulators of T cell activation, *Nature Immunology*, 6:472–480, 2005.

[29] Shen, F., et al., Identification of common transcriptional regulatory elements in interleukin-17 target genes, *J. Biol. Chem.*, 281:24138–24148, 2006.

[30] Smith-Garvin, J.E., Koretzky, G.A., Jordan, M.S., T cell activation, *Annual Review of Immunology*, 27:591–619, 2009.

[31] Sugimoto, N., et al., Foxp3-dependent and -independent molecules specific for CD25+CD4+ natural regulatory T cells revealed by DNA microarray analysis, *International Immunology*, 18:1197–1209, 2006.

[32] Suzuki, H., et al., The transcriptional network that controls growth arrest and differentiation in a human myeloid leukemia cell line, *Nature Genetics*, 41:553–562, 2009.

[33] Takaki, H., et al., STAT6 inhibits TGF-beta1-mediated Foxp3 induction through direct binding to the Foxp3 promoter, which is reverted by retinoic acid receptor, *J. Biol. Chem.*, 283:14955–14962, 2008.

[34] Thompson, C.B., et al., Cis-acting sequences required for inducible interleukin-2 enhancer function bind a novel Ets-related protein, Elf-1, *Molecular and Cellular Biology*, 12:1043–1053, 1992.

[35] Tone, Y. et al., Smad3 and NFAT cooperate to induce Foxp3 expression through its enhancer, *Nature Immunology*, 9:194–202, 2008.

[36] Venuprasad, K., et al., The E3 ubiquitin ligase itch regulates expression of transcription factor Foxp3 and airway inflammation by enhancing the function of transcription factor TIEG1, *Nature Immunology*, 9:245–253, 2008.

[37] Yao, Z., et al., Nonredundant roles for Stat5a/b in directly regulating Foxp3, *Blood*, 109:4368–4375, 2007.

[38] Zhang, F., Meng, G., Strober, W., Interactions among the transcription factors Runx, RORgammat and Foxp3 regulate the differentiation of interleukin 17-producing T cells, *Nature Immunology*, 9:1297–1306, 2008.

[39] Zheng, S.G., et al., Generation ex vivo of TGF-beta-producing regulatory T cells from CD4+CD25- precursors, *Immunology* , 169:4183–4189, 2002.

[40] Zheng, Y., et al., Genome-wide analysis of Foxp3 target genes in developing and mature regulatory T cells, *Nature*, 445:936–940, 2007.

[41] Zhu, J., et al., GATA-3 promotes Th2 responses through three different mechanisms: induction of Th2 cytokine production, selective growth of Th2 cells and inhibition of Th1 cell-specific factors, *Cell Research*, 16:3–10, 2006.

ANNOTATING GENE FUNCTIONS WITH INTEGRATIVE SPECTRAL CLUSTERING ON MICROARRAY EXPRESSIONS AND SEQUENCES

LIMIN LI[1]

limin@hkusua.hku.hk

WAI-KI CHING[1]

wching@hkusua.hku.hk

MOTOKI SHIGA[2]

shiga@kuicr.kyoto-u.ac.jp

HIROSHI MAMITSUKA[2]

mami@kuicr.kyoto-u.ac.jp

[1] *Advanced Modeling and Applied Computing Laboratory, Department of Mathematics, The University of Hong Kong, Pokfulam Road, Hong Kong*
[2] *Bioinformatics Center, Institute for Chemical Research, Kyoto University, Gokasho, Uji 611-0011, Japan*

Annotating genes is a fundamental issue in the post-genomic era. A typical procedure for this issue is first clustering genes by their features and then assigning functions of unknown genes by using known genes in the same cluster. A lot of genomic information are available for this issue, but two major types of data which can be measured for any gene are microarray expressions and sequences, both of which however have their own flaws. Thus a natural and promising approach for gene annotation is to integrate these two data sources, especially in terms of their costs to be optimized in clustering. We develop an efficient gene annotation method with three steps containing spectral clustering over the integrated cost, based on the idea of network modularity. We rigorously examined the performance of our proposed method from three different viewpoints. All experimental results indicate the performance advantage of our method over possible clustering/classification-based approaches of gene function annotation, using expressions and/or sequences.

Keywords: Gene function annotation; spectral clustering; network modularity; integrative clustering.

1. Introduction

Proteins play important roles in a lot of biological functions in a cell and thus functional annotation of genes is a fundamental problem in the post-genomic era. Even for yeast, one of the most well-studied organisms, about one-fourth of all genes still remain uncharacterized [20]. We address the issue of predicting gene functions based on clustering (or grouping) genes into modules [1, 41]. That is, genes are clustered and then functions of an unknown gene are predicted by using functionally known genes in the same cluster (or module).

Various types of data have been used for predicting gene functions in the literature, including sequences, expressions, protein-protein interactions and phylogenetic profiles, but basically only sequences and microarray expressions are the information which can be measured for any gene. However, a problem is that each of these two information sources has its own flaws. Sequence similarity is, in most cases, correlated to functional similarity, but there exist exceptions. In fact it is reported that even highly aligned sequences have totally different functions in some cases [25]. Microarray expressions are weak at data quality: bad probes can be contained and elements in an expression matrix can be corrupted.

Thus, integrating these two data sources is a natural and promising direction to achieve a better performance for the problem of annotating functions of any gene. However, these two data types are different, and integrating sequences with expressions is not simple. An expression profile is a numerical vector, being uniquely mapped in a space of microarray experiments, which results in that microarray expressions are, in machine learning, *structured data* with examples in rows and features of examples in columns. On the other hand, for sequences, similarities are useful for gene clustering, meaning that this data can be an association (or affinity) matrix between sequences, resulting in a weighted network or graph, which is, in machine learning, *unstructured data.* A possible approach for combining these two data types is "data transformation" between each other. For example, expression profiles can be converted into an association matrix by computing correlations between all possible pairs of expression profiles. Then, we can integrate the expression association with the sequence association to generate a total association matrix, which can be a gene network with association weights on edges. In fact, network-based function prediction has been pronounced recently [27]. The reverse way is also possible. That is, a sequence profile can be generated from each sequence and added to the expression profile of the same gene. Once we make such "joint profiles", a lot of techniques including k-means and hierarchical clustering can be employed for clustering them. Furthermore if we fix the set of gene functions beforehand, the problem can be a multi-class classification problem, and in fact classification techniques such as support vector machine (SVM) have been already applied to gene annotation [6, 7] though the data is not necessarily confined to sequences and expressions. We note that this type of data transformation has problems: First, significant information might be lost, being easily implied by the fact that we cannot reproduce the original data from the transformed data. Another doubt can be casted whether a simple addition of two datasets in the above manner would succeed or not.

Clustering numerical vectors usually results in minimizing some cost function, such as a dissimilarity cost function [17]. Similarly clustering over associations, i.e. a gene network, can be an optimization problem which minimizes some criterion, such as normalized cut over a network [28]. Thus we propose an integrative clustering method which minimizes the total cost obtained by linearly weighting the two clustering costs, which are derived from the two different types of datasets. We emphasize that our cost optimization approach is more natural and adequate than the above data transformation approach, because we can use original data directly without losing any information.

In computer science, a close problem setting of ours is constrained clustering [37, 38], where examples of numerical vectors are clustered with two constraints:"must-link", in which two examples must be in the same cluster, and "cannot-link", in which two examples must be in different clusters. A more similar problem setting is semi-supervised clustering [3], and a typical approach is based on a hidden Markov random field (HMRF) in which variables (or nodes) on a random field are constrained with each other and probabilistic parameters attached at nodes are estimated from examples [3]. However, HMRF has a large number of probability parameters which need a lot of computational cost to estimate. Furthermore, this optimization is likely to be affected by initial values, implying that a very large number of trials are required to find a relatively favorable result. In fact, the most related work to us must be [29], which uses a new hidden random field to be optimized by an EM (Expectation Maximization) algorithm and suffered from the problems as written in the above for HMRF.

Our method, which is based on *spectral clustering* [23, 36], has a clear advantage over the random field-based methods in computational efficiency because the optimization problem is solved by time-efficient matrix computation. In addition, the initial parameter problem of random fields is far relaxed or solved. We further emphasize that spectral clustering is a high-lighted approach in the current machine learning literature as well as a *de facto* standard approach in modern graph partitioning. Furthermore for the clustering criterion to be optimized, our method focuses on the idea of *network modularity* [21, 22]. This is a well-recognized, important network property as well as small-world phenomena [39], scale-free property [2] and self-similarity [32], which are all common to a lot of modern network-shaped data such as world wide web and various biological networks. In this light, our proposed method is a computationally original as well as powerful approach for the gene annotation problem, which is very significant in the post-genome era.

Our whole scheme for gene annotation has three steps: First, from sequence information we generate a gene network in terms of sequence similarity by using a pairwise sequence alignment algorithm. We then do clustering this network, combining with microarray expressions, by using the proposed spectral clustering method based on the idea of network modularity. Finally we assign functions of unknown genes by using the clusters, which were generated by the preceding step.

We evaluated the performance of our scheme in the following three experiments: 1) We first checked the effectiveness of our way of combination of gene expressions and sequences by checking the overlap between predicted clusters and standard functional clusters. 2) We then examined whether our approach of predicting gene functions can outperform other various approaches, particularly simpler data-transformation-based methods, including network-based, k-means-based and SVM-based approaches. 3) We finally validated resultant annotation on unknown genes in Saccharomyces Genome Database (SGD) [20] by using the information of Comprehensive Yeast Genome Database (CYGD) [14].

This paper is organized as follows: Section 2 shows our proposed method which has three steps: 1) generating a gene network by sequence similarity, 2) clustering nodes of a gene network on sequence similarities with microarray expressions of genes, each of which corresponds to a node of the network, and 3) predicting functions of unknown genes by gene functions corresponding to clusters generated by the preceding step. The second step of our method is based on our past work [30], where we proposed a new, general method for clustering numerical vectors considering their network constraints, and verified the performance by basically using a variety of synthetic data. On the other hand, this paper focuses on the problem of annotating gene functions, particularly using the data of microarray expressions and a gene network on sequence similarities. In fact our method in this paper has three steps, and only the second step is related with [30]. Furthermore, we extensively examined the performance of our method by using real data of gene expressions and sequence similarities. We emphasize that these points make this paper significantly different from [30]. Section 3 shows our experimental results which can be mainly divided into four parts: 1) generating a new gene network, 2) examining our integrative clustering method, 3) checking the predictive performance of our entire method and 4) annotating unknown yeast genes. Section 4 summarizes the proposed approach and discusses the future work on this method.

2. Proposed Method

Figure 1 shows our entire framework with three steps for predicting gene functions based on integrative spectral clustering. We describe the detail of each step in this section.

2.1. *Notation*

Our method has three inputs. The first input is a table of gene expression profiles. Let X be an input profile table, i.e. $X = [x_1, \ldots, x_N]$, where x_i ($= [x_{i1}, \ldots, x_{ip}]^T$) is an expression profile for gene i and x_{ij} is an expression value of gene i in sample j. Let $\tilde{x}_i$ be the normalized vector of x_i satisfying with $\tilde{x}_i^T \tilde{x}_i = 1$, and $\tilde{X}$ be the normalized matrix of X ($\tilde{X} = [\tilde{x}_1, \ldots, \tilde{x}_N]$). Let Y be the inner-product of $\tilde{X}$ ($Y = \tilde{X}^T \tilde{X}$). The second input of our method is a gene network, having a gene as a node. For a given network, let w_{ij} be a numerical weight between nodes i and j, taking a range between zero and one. If an input network is a binary network, w_{ij} takes one if there is an edge between i and j; otherwise zero. Let W be a matrix with w_{ij} for element (i, j). Let $d_i = \sum_{j=1}^{N} w_{ij}$ and $D := d^T d$ where $d := (d_1, \cdots, d_N)$. Let D_d be a $N \times N$ diagonal matrix whose (i, i) entry is d_i. The third input of our method is K, which is the number of clusters.

Let z_{ki} be a binary cluster indicator taking one if gene i is in cluster k; otherwise zero. Let z_k be a vector of the k-th cluster assignment: $z_k = [z_{k1}, \ldots, z_{kN}]^T$, and Z be a matrix of cluster assignments ($Z = [z_1, \ldots, z_K]$). Let μ_k be the center of cluster k, and μ be a set of cluster centers ($\mu = [\mu_1, \ldots, \mu_K]$). Let $\mathrm{tr}(A)$ be the trace of

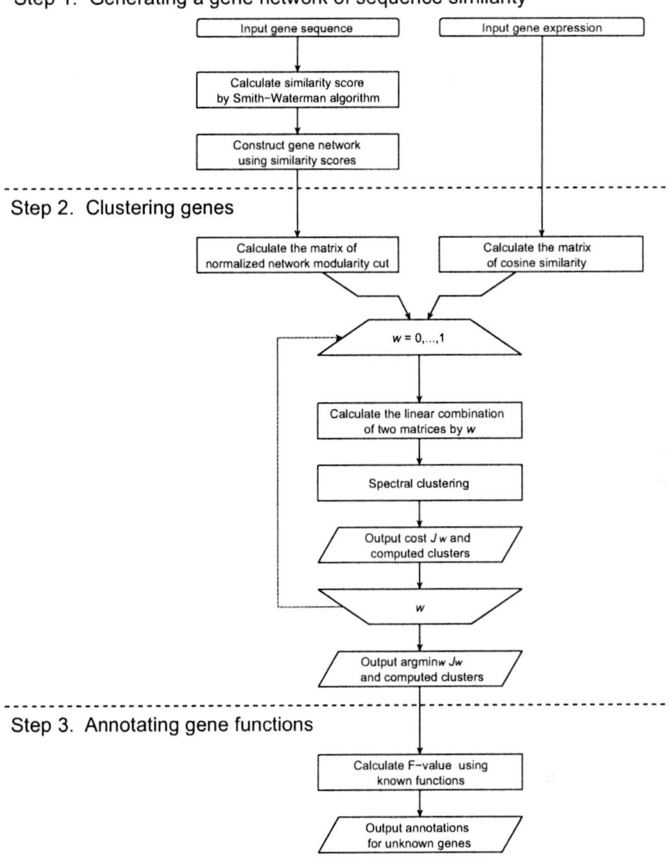

Fig. 1: Entire procedure of our method of annotating gene functions.

matrix A $(\mathrm{tr}(A) = \sum_{i=1}^{n} A_{ii})$. Let $L(z_k, z_{k'})$ be the number of edges between clusters k and k', and L be the number of edges in a network: $L = L(\mathbf{1}, \mathbf{1})$, where $\mathbf{1}$ be the N-dimensional column vector with all elements of one. Let N_C be the number of genes in cluster C, and $N_{C,F}$ be the number of genes which are in both clusters C and F.

2.2. *Step 1: Generating a Gene Network of Sequence Similarities*

As shown in our notation of w_{ij}, our framework can handle a real-valued association matrix or a weighted network, but we use a binary network for sequence similarity in our experiments because of computational efficiency and network reliability. We first align all pairs of sequences using the Smith-Waterman algorithm [31] and obtain a list of sequence pairs with similarity p-values, which are a measure indicating how likely each pairwise alignment can be arised by chance. That is, the smaller the similarity p-value is, the more similar the sequence pair is. As the sequence similarity is correlated to the functional similarity, two genes with a low p-value is supposed to share the same function. We place a cut-off value for the p-values to generate a binary affinity matrix between genes, i.e. a gene network. Concretely, a network is generated by drawing an edge between two nodes if the p-value of the pair corresponding to the two nodes is lower than some specified cut-off value. We call this network a *sequence network*.

To find a better cut-off value, we use some standard functional clusters, to which known genes are assigned correctly. We then change cut-off value v against similarity p-values and compute precision, Prec_v, as follows:

$$\mathrm{Prec}_v = \frac{B_v}{A_v},$$

where A_v is the number of gene pairs with similarity p-values less than v, and B_v is the number of gene pairs with similarity p-values less than v and with shared functions. Note that $B_v \subset A_v$. We select an appropriate cut-off value to generate a large sequence network with a high precision and use this cut-off value throughout all experiments.

2.3. *Step 2: Clustering Genes by Combining a Sequence Network with Gene Expressions*

We first define two clustering costs: the cost of clustering numerical vectors and that of clustering nodes over a sequence network. We then linearly combine these two costs and attempt to optimize the total cost by spectral clustering, which transforms the optimization problem into an eigenvalue problem using Lagrange multipliers.

2.3.1. *Two Clustering Costs*

The cost of clustering numerical vectors, i.e. microarray expression profiles, can be defined by the k-means clustering algorithm as follows:

$$J_{num}(X, Z; \mu) = \frac{1}{N} \sum_{k=1}^{K} \sum_{i \mid z_{ki}=1} \mathbf{Dist}(x_i, \mu_k), \tag{1}$$

where $\mathbf{Dist}(x_i, \mu_k)$ is a distance between numerical vector (expression profile) x_i and cluster center μ_k, borrowing the idea of cosine similarity:

$$\mathbf{Dist}(x_i, \mu_k) = \frac{1}{2}\left(1 - \frac{x_i^T \mu_k}{\sqrt{x_i^T x_i}}\right), \tag{2}$$

and cluster center μ_k can be written as follows:

$$\mu_k = \frac{1}{|z_k|} \sum_{j \in z_k} \tilde{x}_j.$$

By using Eq. (2), Eq. (1) can be transformed into the trace optimization problem as follows:

$$
\begin{aligned}
J_{num}(\tilde{X}, Z; \mu) &= \frac{1}{2N} \sum_{k=1}^{K} \sum_{i \in z_k} \left(1 - \tilde{x}_i^T \mu_k\right) \\
&= \frac{1}{2N} \sum_{k=1}^{K} \sum_{i \in z_k} \left(1 - \tilde{x}_i^T \frac{1}{|z_k|} \sum_{j \in z_k} \tilde{x}_j\right) \\
&= \frac{1}{2} - \frac{1}{2N} \sum_{k=1}^{K} \frac{1}{|z_k|} \sum_{i \in z_k} \sum_{j \in z_k} \tilde{x}_i^T \tilde{x}_j \\
&= \frac{1}{2} - \mathrm{tr}\left(\frac{Z^T (2N)^{-1} Y Z}{Z^T Z}\right).
\end{aligned}
\tag{3}
$$

Graph partitioning needs some criterion for grouping clusters. The simplest criterion is *graph min-cut*, which minimizes the number of inter-cluster edges, i.e. the number of total edges from a cluster to other clusters, but is likely to generate small clusters which in some cases correspond to outliers. To overcome this shortcoming of graph min-cut, graph min-cut has been normalized in some ways such as *ratio-cut* [8, 15], *min-max cut* [9] and *normalized cut* [28]. In particular, a *de facto* standard criterion for k-way graph partitioning is normalized cut, which is obtained by dividing graph min-cut by the number of edges at each cluster.

In this paper, as the graph cut criterion, we employ the idea of *network modularity* [13, 21, 22], which has been paid a lot of attentions as an important network property in the recent network science. In fact, this property has been found in modern network-shaped data such as metabolic networks [12, 24] and is closely related with network clustering. The network modularity is defined as follows:

$$
Q(W, Z) = \sum_{k=1}^{K} \left\{ \frac{L(z_k, z_k)}{L} - \left(\frac{L(z_k, 1)}{L}\right)^2 \right\}.
$$

As well as graph min-cut, the issue of the cluster size, i.e. the outlier problem, is unconsidered by the original network modularity. Thus we define the new graph-cut criterion, *normalized network modularity cut*, and the cost of graph partitioning (or clustering genes in a given sequence network) of this cut can be given as follows:

$$
J_{net}(W, Z) = \sum_{k=1}^{K} \frac{N}{\sum_i z_{ki}} \left\{ \left(\frac{L(z_k, 1)}{L}\right)^2 - \frac{L(z_k, z_k)}{L} \right\}
$$

We can further transformed the new clustering criterion into a trace optimization framework, borrowing the idea of [40] on the (original) network modularity, as follows:

$$
\begin{aligned}
J_{net}(W, Z) &= \sum_{k=1}^{K} \frac{N}{\sum_i z_{ki}} \left\{ \left(\frac{\sum_{i,j} w_{ij} z_{ki}}{L} \right)^2 - \frac{\sum_{i,j} w_{ij} z_{ki} z_{kj}}{L} \right\} \\
&= \sum_{k=1}^{K} \frac{N}{\sum_i z_{ki}} \left\{ \left(\frac{d^T z_k}{L} \right)^2 - \frac{z_k^T W z_k}{L} \right\} \\
&= \sum_{k=1}^{K} \frac{N}{\sum_i z_{ki}} \left\{ \frac{z_k^T d d^T z_k}{L^2} - \frac{z_k^T W z_k}{L} \right\} \\
&= \mathrm{tr} \left(\frac{Z^T N(\frac{1}{L^2} D - \frac{1}{L} W) Z}{Z^T Z} \right)
\end{aligned}
\tag{4}
$$

2.3.2. *Optimizing Total Cost*

From the obtained two costs in the preceding section, one can easily see that these two costs share the same denominator. Thus, we combine the above two costs in trace optimization linearly by using weight ω, which is a control parameter taking a real value in the range from zero to one, for balancing the two costs:

$$
\begin{aligned}
J_{total}(\tilde{X}, W, Z) &= \omega J_{net}(W, Z) + (1 - \omega) J_{num}(\tilde{X}, Z) \\
&= \mathrm{tr} \left(\frac{Z^T (\frac{\omega N}{L^2} D - \frac{\omega N}{L} W - \frac{1-\omega}{2N} Y) Z}{Z^T Z} \right)
\end{aligned}
$$

We can easily see that only microarray expressions are considered for $\omega=0$ (denoted by ω_0 in our experiments), while only a given network is used for $\omega=1$ (denoted by ω_1 in our experiments). Our strategy is to solve this trace optimization problem under the constraint of the indicator matrix Z. We first, by relaxing the binary cluster indicators into real values, formulate the clustering problem as the following optimization problem:

$$
\text{minimize} \ \ \mathrm{tr} \left(Z^T (\frac{\omega N}{L^2} D - \frac{\omega N}{L} W - \frac{1-\omega}{2N} Y) Z \right) \quad \text{s.t.} \quad Z^T Z = I_K,
\tag{5}
$$

where I_K is the identity matrix of size K. In our experiments, because of practical stability as implied by [40], we used $Z^T D_d Z = I_K$ instead of $Z^T Z = I_K$. Then this optimization problem can be transformed into the following eigenvalue problem by using Lagrange multipliers:

$$
M_\omega Z = Z \Lambda,
$$

where $M_\omega = D_d^{-\frac{1}{2}}(\frac{\omega N}{L^2} D - \frac{\omega N}{L} W - \frac{1-\omega}{2N} Y) D_d^{-\frac{1}{2}}$. Let U $(= [u_1, \ldots, u_{K-1}] \in \mathbb{R}^{N \times (K-1)})$ denote the matrix containing $K - 1$ eigenvectors of M_ω, corresponding to the first $K - 1$ smallest eigenvalues. Finally k-means can be applied to the normalized N row vectors of U to obtain gene clusters to be output.

This flow of our method exactly follows spectral clustering, a typical k-way graph partitioning approach, which is a trace optimization with constraints. For example,

---

Input : X, W, K, $Z^{(0)}$, $\mu^{(0)}$
Output : Z

1: $\tilde{X} \leftarrow X/\sqrt{X^T X}$
2: $Y \leftarrow \tilde{X}^T \tilde{X}$
3: Compute D_d and D from W
4: **for** $\omega = 0$ to 1 **do**
5: $\quad M_\omega \leftarrow D_d^{-\frac{1}{2}} (\frac{\omega N}{L^2} D - \frac{\omega N}{L} W - \frac{1-\omega}{2N} Y) D_d^{-\frac{1}{2}}$.
6: $\quad$ Compute U of M_ω.
7: $\quad$ Normalize U into $\tilde{U}$.
8: $\quad [Z_\omega, \mu_\omega, J_\omega] \leftarrow$ **k-means**$(\tilde{U}, K, Z^{(0)}, \mu^{(0)})$
9: **end for**
10: $\omega_{\min} \leftarrow \arg\min_\omega J_\omega$
11: $Z \leftarrow Z_{\omega_{\min}}$

Fig. 2: Pseudocode of the clustering algorithm in Step 2.

the cost by the normalized cut can be the following trace:

$$\sum_k \frac{L(z_k, \mathbf{1} - z_k)}{L(z_k, \mathbf{1})} = \sum_k \frac{z_k^T (D_d - W) z_k}{z_k^T D_d z_k}$$
$$= \mathrm{tr}\left(\frac{Z^T (D_d - W) Z}{Z^T D_d Z}\right). \tag{6}$$

Graph partitioning with normalized cut can be then the following optimization problem:

$$\text{minimize} \quad \mathrm{tr}(Z^T (D_d - W) Z) \quad \text{s.t.} \quad Z^T D_d Z = I_K \tag{7}$$

We note that this trace optimization manner is basically the same as that of ours. However precisely, one can see the difference between our optimization problem and the above (normalized cut-based) problem by examining Eq. (5) and Eq. (7). That is, the above problem can be the so-called *general* eigenvalue problem. Thus an important point is that in principle the normalized cut (as well as the original network modularity) cannot be used to combine numerical vectors with a network, although the normalized cut is a popular criterion in graph partitioning. More precisely, one can see the difference by examining the denominators of the costs. That is, the denominator of the cost of numerical vectors is $Z^T Z$ (Eq. (3)), which is the same as that by normalized network modularity (Eq. (4)) but not that by normalized cut (Eq. (6)). This indicates that numerical vectors and a network can be combined by using the normalized network modularity cut which we propose.

A further key feature of our integrative approach is that we can select the optimal value of parameter ω, which takes a balance between the cost of numerical vectors and that of a network, by choosing the ω which can give the minimum cost in the final k-means step in the eigenspace. This cost minimization is to select the eigenspace in which the clusters are separated the most from each other, and this

---

Input : X, K, $Z^{(0)}$, $\mu^{(0)}$
Output : Z, μ, J

k-means $(X, K, Z^{(0)}, \mu^{(0)})$
1: $Z \leftarrow Z^{(0)}$, $\mu \leftarrow \mu^{(0)}$
2: $\tilde{X} \leftarrow X/\sqrt{X^T X}$
3: **while** $J(\tilde{X}, Z; \mu)$ is not converged **do**
4:   $\mu_k^{(t+1)} \leftarrow \dfrac{1}{|Z_k^{(t)}|} \sum_{j \in Z_k^{(t)}} \tilde{x}_j \quad (k = 1, \cdots, K)$
5:   $Z^{(t+1)} \leftarrow \arg\min_Z J(\tilde{X}, Z; \mu^{(t+1)})$
6:   $Z \leftarrow Z^{(t+1)}$, $\mu \leftarrow \mu^{(t+1)}$, $J \leftarrow J(\tilde{X}, Z; \mu)$
7: **end while**

Fig. 3: Pseudocode of k-means.

eigenspace must be the best for clustering genes. We use this optimal ω to have gene clusters which are used for predicting functions of genes.

Figure 2 shows a pseudocode of our entire algorithm of integrative clustering with normalized network modularity, and Figure 3 shows a pseudocode of k-means which is used in Figure 2.

2.4. *Step 3: Predicting Functions of Unknown Genes*

We first assume that we have a set of standard functions (or functional modules) for which known genes are already assigned correctly. We note that any set can be used for our method. In this setting, our annotation strategy, which assumes that genes in the same cluster share the same function, has two steps. 1) We first examine the relevance between each of computed clusters and each of standard functions by using some measure. 2) We then, for each cluster, rank the functions according to the relevance, and the most relevant function is assigned to genes in this cluster as their functions. The problem is how to compute/score the relevance between a computationally obtained cluster and each of standard functions.

We first introduce our default measure, which we call **F**-value (standing for frequency-value) and is computed for function F as follows:

$$\mathbf{F}\text{-value} = \frac{\#\text{ genes categorized in } F}{\#\text{ genes}}.$$

We then show a more popular scoring measure: **P**-value between computed cluster C and standard function F:

$$\mathbf{P}\text{-value} = 1 - \sum_{i=0}^{N_{C,F}-1} \frac{\dbinom{N_F}{i} \dbinom{N - N_F}{N_C - i}}{\dbinom{N}{N_C}}.$$

This value measures the probability that C is enriched by F by chance. More concretely, if **P**-value is close to zero, the probability that a gene in the clusters

Table 1: Microarray expression data summary.

Name	#observations	#missing values	Reference
Brem	103	2172	[5]
Gasch	140	1462	[11]
Hughes	300	241	[18]
Spellman	57	2110	[33]
Storey	262	6946	[34]
Yvert	180	4405	[42]

is chosen by chance will be close to zero. That is, major functions of a cluster should have smaller **P**-values. The **P**-value has been well used in annotating tools, such as GOTermFinder [4]. However, we note that **P**-value, which originates from the hypergeometric distribution, heavily depends upon the number of standard functions. For example, even if all genes in a cluster have the same function, the corresponding **P**-value might not be small enough if the number of genes having this function is much larger than that of the cluster.

3. Experiments

3.1. *Data*

We focused on genes of *Saccharomyces Cerevisiae*. We used a list of sequence similarities which are already computed and stored in the SGD database [20]. This list has gene pairs with similarity p-values which are less than 0.01. We used six gene expression datasets shown in Table 1 to validate the performance of our clustering method. Missing values in these six data sets were all imputed by a method based on k-nearest neighbors [35] with $k = 10$.

We used GO (Gene Ontology) to generate a (gold-)standard dataset of fifteen gene functions, which was used through our experiments for some purposes, particularly for evaluating the clusters generated by our method. The manner of generating these standard functions is described in Appendix A. The GO was obtained from http://www.geneontology.org/ontology/gene_ontology.obo.

3.2. *Examining Gene Networks on Sequence Similarities*

We examined sequence similarities to generate a sequence network, following the manner of Step 1 of our method.

3.2.1. *Generating a Sequence Network*

Figure 4 shows the plots of $Prec_v$ which were obtained by changing cut-off value v against p-values, where we used the fifteen standard functions generated in Section 3.1. From the figure, we can see that genes with high similarities in sequences

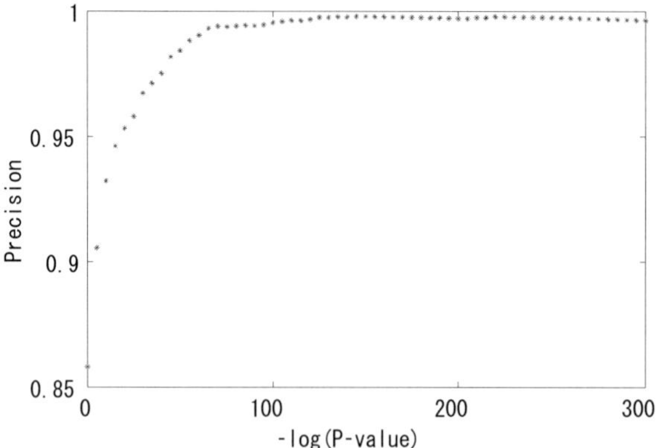

Fig. 4: Prec_v against sequence similarity p-values.

Table 2: Gene sets (in which a node is a gene) used in Sections 3.3 to 3.5.

Sets	# nodes	# edges	# nodes in LCC	# edges in LCC	Section used
I	2,025	7,972	487	4,450	3.3 & 3.4
II	2,305	9,710	582	5,686	3.4
III	2,903	12,642	834	8,229	3.5

are very likely to be in the same functional clusters (or share the same functions). From the figure, we chose the cut-off of 1.0e-5, corresponding to Prec_v of 0.9056, since the network at the next point by Prec_v of 0.9324 had only 90 nodes in the largest connected component (LCC) while the network at the cut-off of 1.0e-5 had 487 nodes in the LCC. We here note that a sequence network is generated from sequence similarities, and the fifteen standard functions were just used to decide the cut-off value which can give a network with some moderate size. Even if we use another standard set of functions, the cut-off value will be 1.0e-5 due to the network size of LCC, implying that the same sequence network will be generated, regardless of given standard functions.

At this cut-off value of 1.0e-5, we generated three different types of networks, i.e. gene sets: 1) Set I: was genes which can be categorized into one or more of the fifteen standard clusters, 2) Set II: was genes which can be labeled by any GO terms including the fifteen standard clusters and 3) Set III: was all genes including unknown genes or gene unlabeled with any GO terms. Table 2 shows the statistics of these three sets. Note that Set I $\subset$ Set II $\subset$ Set III. We used the LCC only throughout this paper.

Table 3: Normalized mutual information (R:Ratio cut, N: Normalized cut).

Dataset	ω_0	ω_1	ω^*	p-value (ω^* vs. ω_1)	R	N
(a) $K{=}10$						
Brem	0.0676	0.2694	**0.2949**	1.16e-13		
Gasch	0.0595	0.2733	**0.2892**	4.86e-09		
Hughes	0.0549	0.2641	**0.2862**	7.10e-14	0.2806	0.2799
Spellman	0.0417	0.2628	**0.3011**	3.55e-23		
Storey	0.0586	0.2715	**0.2955**	6.97e-13		
Yvert	0.0602	0.2662	**0.3019**	3.63e-21		
(b) $K{=}15$						
Brem	0.0766	0.3107	**0.3245**	9.90e-14		
Gasch	0.0839	0.3104	**0.3263**	6.37e-16		
Hughes	0.0819	0.3090	**0.3267**	1.45e-15	0.3185	0.3132
Spellman	0.0642	0.3117	**0.3309**	3.26e-16		
Storey	0.0953	0.3109	**0.3283**	1.96e-15		
Yvert	0.0866	0.3097	**0.3252**	4.63e-13		
(c) $K{=}20$						
Brem	0.0968	0.3493	**0.3679**	2.74e-17		
Gasch	0.0990	0.3518	**0.3664**	2.92e-16		
Hughes	0.0974	0.3536	**0.3639**	1.23e-11	0.3517	0.3418
Spellman	0.0746	0.3504	**0.3661**	6.75e-13		
Storey	0.1161	0.3539	**0.3665**	1.52e-17		
Yvert	0.1096	0.3497	**0.3660**	8.83e-16		

3.3. *Evaluating the Performance of Integrative Clustering Part*

We examined the performance of Step 2 of our method using microarray data and a sequence network, which was generated in the preceding section. We note that Step 3 of our method was not used in this experiment.

3.3.1. *Competing Methods*

We examined the performance of our clustering method (Step 2) with changing ω from zero (ω_0) to one (ω_1) at the interval of 0.1. Out of the eleven results obtained in this manner, we chose the optimal ω^* which gave the minimum cost. We compared the clustering performance at ω^* with those of ω_0 and ω_1. Furthermore, we compared the performance of our clustering method with two graph clustering methods using the sequence network only, i.e. graph clustering with ratio cut [8, 15] and normalized cut [28].

3.3.2. *Data and Procedure*

We used Set I and six gene expression datasets in Table 1. As described in Section 2.3, our clustering algorithm uses k-means at the final step, and the results by

k-means depend on initial cluster centers. Thus we performed 100 trials with different random initial centers for each ω and averaged over the 100 runs to make a fair comparison.

3.3.3. *Evaluation Criterion*

We checked the performance of our method by comparing resultant clusters with gold-standard clusters, for which we used fifteen GO functions we generated in Section 3.1. For evaluation criterion, we used normalized mutual information (NMI) which is defined in information theory as a quantity to measure the amount of information shared between two random variables. NMI is defined by the following formula:

$$\text{NMI} = \frac{I(V, V')}{\sqrt{H(V) \cdot H(V')}} \tag{8}$$

where V and V' are the calculated clusters and standard clusters, respectively. $I(V, V')$ is the mutual information between V and V', and $H(V)$ and $H(V')$ are the entropy of V and V'. This can be estimated by the following formula [43]:

$$\text{NMI} = \frac{\sum_{C,C'} N_{C,C'} \log(\frac{N \cdot N_{C,C'}}{N_C N_{C'}})}{\sqrt{(\sum_C N_C \log \frac{N_C}{N})(\sum_{C'} N_{C'} \log \frac{N_{C'}}{N})}} \tag{9}$$

where C is a computed cluster and C' is a standard function. NMI takes a value ranging from 0 to 1, and the closer to one it is, the more similar to standard clusters the computed clusters are. We note that NMI is the most typical measure for checking the performance of a clustering method.

3.3.4. *Performance Comparison*

Table 3 shows the averaged NMI for the six datasets at $K = 10$, 15 and 20. For all cases of this table, the averaged NMI at ω^* was slightly better than that at ω_1, which was clearly better than that at ω_0. Since the difference ω^* and ω_1 looked small, we then applied paired t-test between them to confirm the statistical significance in the difference between them, and p-values of this test are also reported in the fifth column. Table 3 further shows the averaged NMI obtained by using ratio cut and normalized cut, which were almost the same level as those by ω_1. All these results imply that the optimal ω outperformed ω_0 and ω_1, meaning that combining the two different datasets by our method was significantly effective for gene clustering. As ω^* outperformed others for all K, we used $K{=}10$ in Sections 3.4 and 3.5.

To further evaluate the effectiveness of the cost function, we then focused on Spellman, which was with the highest NMI and the smallest p-value in Table 3 (a), and checked the consistency between the NMI and the cost function. Figure 5 shows the costs (in the top row) and NMI (in the bottom row) with varying ω from zero to one for the three cases of K: 10, 15 and 20. Blue points in each figure show 100

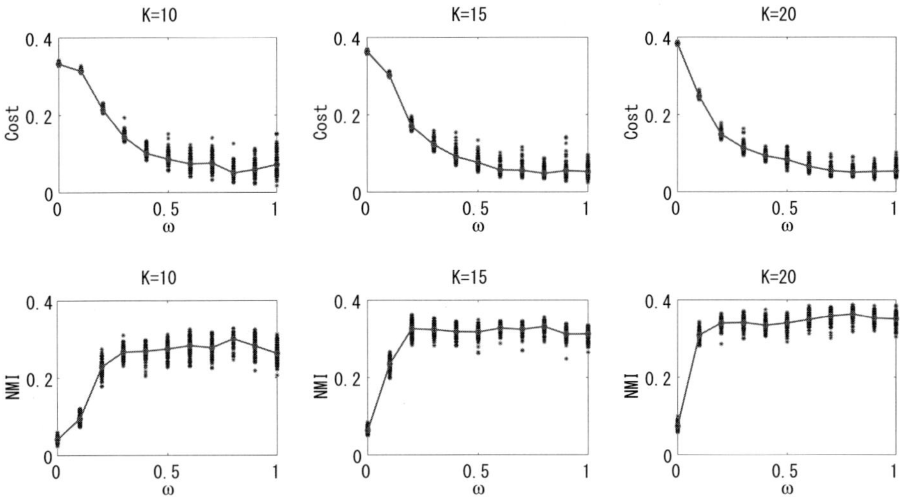

Fig. 5: Top: total costs of K=10, 15 and 20; Bottom: NMI of K=10, 15 and 20.

values corresponding to 100 repeated runs at each trial, and a red curve shows the average over the 100 runs. One can see that for all three cases of K, the ω with the minimum cost resulted in the highest NMI, meaning that our strategy of using the cost function to find $\omega*$ was useful. Another finding was that the optimal ω in the cost was 0.8 for all three cases, and then hereafter we set ω^* at 0.8.

3.4. *Evaluating the Predictive Performance of Our Entire Method*

We examined the performance of our method including Step 3. Here let us make clear the difference between Sections 3.3 and 3.4. In Section 3.3, genes were clustered and this grouping was evaluated by comparing with standard functions (clusters). On the other hand, in Section 3.4, as will be described later, cross-validation was performed, meaning that genes were divided into training and testing, and once clustered, functions of each gene in the test set were predicted by the functions of training genes in the cluster into which the test gene was fallen. Thus one can say that clustering performance was checked in Section 3.3, while predictive performance was examined in Section 3.4.

3.4.1. *Competing Methods*

We first examined three parameter settings of our method: ω^*, ω_0 and ω_1 with **F**-value and **P**-value. We then considered a simple approach, Random, with the following procedure: We first assumed ten clusters and, for each cluster, randomly assigned a GO term with a score randomly taking a value between zero and one. We then randomly assigned one of ten cluster labels to each gene as a predicted cluster label.

Next we examined the performance of network-based approaches. In fact, ω_1 of our method is equal to a network-based clustering of normalized network modularity cut over a sequence network. We, from gene expressions, first generated a network, which we call an *expression network*, by drawing an edge between two genes if the cosine similarity in expressions of the two genes is larger than some prefixed cut-off value θ. We then performed clustering nodes of the following three types of networks: 1) an expression network only (ExpNet), 2) the intersection of an expression network and a sequence network (IntSecNet) and 3) the union of the two networks (UnionNet). Clustering over these three networks was done by spectral clustering with normalized network modularity to make a fair comparison with our method. We tested both **F**-value and **P**-value.

We further examined the performance of "joint profile"-based approaches. We can first generate a feature vector from each gene sequence by using its amino acid compositions, hydrophobicity, normalized van der Waars volume, polarity and polarizability, which are well used in this type of transformation, e.g. [7]. We then examined three types of feature vectors as inputs for k-means: 1) expression profile only (ExpKM), which is basically equal to ω_0, 2) sequence feature only (SeqKM) and 3) their combination (ComKM). We further explored the performance of "joint profile"-based approaches by examining SVM, fixing the cluster set at the fifteen standard clusters, which were generated in Section 3.1. As well, this case, we used the above three networks, and we call the results obtained by 1), 2) and 3), ExpSVM, SeqSVM and ComSVM, respectively. For ExpSVM, we used the cosine similarity kernel to make a fair comparison with our method, and for SeqSVM, we tested three different kernels: linear (SeqLiSVM), polynomial (SeqPoSVM) and Gaussian (SeqGaSVM). The kernel of ComSVM was the simple addition of the cosine similarity kernel of ExpSVM and SeqGaSVM, which was chosen for SeqSVM since in our experiments there were no significant difference in performance among three SeqSVMs, comparing to ω^*. SVM is a binary classifier, and so for fifteen standard functions, we generated fifteen SVMs by repeating training a SVM using genes of each function as positives and others as negatives. Prediction was performed by taking the highest prediction score among the predictions of all fifteen trained SVMs. For SVM, we used Matlab with the QP (Quadratic Programming) option and optimized the regularization parameter by examining a wide variety of possible values.

3.4.2. *Data and Procedure*

We conducted a 100x5-fold cross validation over Set II. That is, we divided all genes into five blocks of roughly equal size, and at each trial four out of five were used for training and the remaining one was for test. We repeated this division 100 times and the results were averaged over the 100 runs. More precisely, in all clustering approaches, each run is that all genes were first clustered, and functions of each gene in the test set were predicted by using only genes in the training dataset. On the other hand, in SVM-based approaches, training and test genes are separately used for training and testing phases, respectively.

We used Set II for training and Set I for testing. However, for SVM-based approaches, we used Set I for training, since in classification, we needed a fixed size of true classes. This implies that a larger number of genes are available in training of clustering than that of classification. The Spellman dataset was used as gene expressions.

3.4.3. *Evaluation Criterion*

We used Set I genes for testing, meaning that all genes in testing can be categorized into one or more fifteen standard functions that we generated in Section 3.1. We note that except SVM, we used a larger number of GO clusters in training.

We evaluated all competing methods in this experiment in a binary classification manner by regarding each pair of a gene and a function as an example. First, for each pair of gene i and function k, we can assign one (positive) if gene i has function k; otherwise zero (negative). This can be a true label of the pair of gene i and function k. Second, a clustering method in this experiment can output a score whether gene i has function k, because **F**-value (or a similar quantitative score) of function k can be computed from the cluster into which gene i was fallen. (We note that similarly SVM-based approaches can also easily output a score for each pair of gene i and function k, because for each function, an SVM was trained.) We then sorted examples (pairs) according to resultant scores and computed AUC (Area Under the ROC (Receiver Operator Characteristic) curve) by using the sum of the rank of negatives. For the detailed procedure of computing AUC in binary classification task, interested readers can consult some work on AUC, e.g. [16, 19].

3.4.4. *Results*

Table 4 shows the AUC of all compared methods. First, the AUC of ω^* was significantly better than those of ω_1 and ω_0, being consistent with Table 3. Naturally the AUC of Random was almost 0.5, being equal to random guessing. For network-based approaches, cut-off θ for cosine similarities is a parameter to change the number of edges in an expression network. In fact, expression networks had 87,538 edges at θ of 0.25 and 13,826 edges at θ of 0.50 while only 1,016 edges at 0.75. This difference changed the AUC of IntSecNet, which was around 0.69 at 0.25 and 0.50 but dropped drastically at 0.75. This is probably because the intersection network at θ of 0.75 was too sparse. However, in ExpNet, the performance was relatively stable through the three values of θ. In UnionNet, AUC clearly raised with increasing θ, implying that this combination worked in prediction. In fact, AUC at θ of 0.75 reached 0.7789 which looked very close to that of ω^*. We then computed paired t-test to examine the significance between these two methods and found that its p-value was 1.0358e-61, implying that the difference in AUC was statistically significant. All results of "joint-profile"-based approaches were much

Table 4: AUC of competing methods.

Method	AUC		
Our method			
ω_0	0.6728		
ω_1	0.7359		
ω^*	**0.7828**		
Simple approach			
Rand	0.4792		
Network-based approaches			
	$\theta{=}0.25$	$\theta{=}0.50$	$\theta{=}0.75$
ExpNet	0.6547	0.6491	0.6502
IntSecNet	0.6934	0.6910	0.6368
UnionNet	0.7051	0.7443	0.7789
k-means-based approaches			
	$k{=}10$	$k{=}15$	$k{=}20$
ExpKM	0.6594	0.6530	0.6619
SeqKM	0.6832	0.6954	0.6837
ComKM	0.6988	0.6856	0.6997
SVM-based approaches			
SeqLiSVM	0.6629		
SeqPoSVM	0.6766		
SeqGaSVM	0.6709		
ExpSVM	0.6724		
ComSVM	0.6827		

lower than AUC of ω^*. In particular, the results of SeqSVM were worse than ω_1 through all three kernels, while ExpSVM was comparable to ω_0, implying that for sequences, a network-based approach, especially UnionNet was more favorable than using sequence features. Another item of note is that there might be some room to improve the performance of SVM-based methods, but the results in Table 4 imply that our method will outperform them even if some improvements are done on the SVM-based methods.

Table 5 shows the AUC of the competing methods when **P**-value was used in prediction. The AUC of each method in this table was much worse than that of the corresponding method of Table 4, indicating that **F**-value was more useful than **P**-value.

Overall, we can conclude that ω^* with **F**-value was the best among competing methods for predicting gene functions, including network-based, k-means-based and SVM-based approaches.

Table 5: AUC of competing methods with **P**-value.

Method	AUC		
Our method			
ω_0	0.5839		
ω_1	0.6951		
ω^*	0.7270		
Network-based approaches			
	$\theta=0.25$	$\theta=0.50$	$\theta=0.75$
ExpNet	0.5405	0.5062	0.5463
IntSecNet	0.5161	0.6153	0.5323
UnionNet	0.6347	0.6907	0.7246
k-means-based approaches			
	$k=10$	$k=15$	$k=20$
ExpKM	0.5644	0.5425	0.5605
SeqKM	0.6123	0.6164	0.6287
ComKM	0.6334	0.6089	0.6369

3.5. *Annotating Yeast Genome*

As we confirmed the performance advantage of our method over a lot of other methods, we applied our method to annotate functions of genes whose functions are still unknown.

3.5.1. *Data and Procedure*

We used Set III and Spellman. We run ω^* with k of 10, 1,000 times, and then predicted gene functions by using the average clustering results over the 1,000 runs and **F**-value.

3.5.2. *Results*

Table 6 shows resultant ten clusters, with uncharacterized genes and GO terms with **F**-values of more than 0.5. Some uncharacterized genes in SGD are already annotated by another database. We validated unknown genes in Table 6 by checking FunCat [26] of CYGD [14]. In particular, we focused on the top three clusters in Table 6 show the analysis on them. In C_1, YLR289w, the only uncharacterized gene in C_1, was annotated to "translation" in FunCat, being the same as the GO term which was ranked at the top by **F**-value in Table 6. In C_2, we found that all genes were annotated by FunCat to either "Modification by phosphorylation, dephosphorylation, autophosphorylation" or "Phosphate metabolism", except only two genes YGL083w and YJL057c, which were uncharacterized in FunCat. These two functional categories were mostly consistent with the predicted function: "Protein amino acid phosphorylation", which was GO:6468. In fact, "Modification by phosphorylation, dephosphorylation, autophosphorylation" and "Phosphate metabolism" were GO:16310 and GO:6796, respectively, and GO:6468 was the only child of GO:16310, which was also the only child of GO:6796. This fact indicates that these functions were firmly related with each other. In C_3, all genes were annotated by FunCat to "tRNA processing" (except two genes: YGL131c and YPR022c), which was a descendant of "transcription", while the predicted function "Regulation of transcription from RNA polymerase II promoter" was also a descendant of "transcription" in GO, meaning that these two functions were also related with each other. Overall these observations imply the preciseness in annotation by our approach.

4. Conclusion and Discussion

We have proposed a new method for predicting gene functions using sequence similarities and microarray expressions. From our thorough experiments, we have shown that our method can optimally combine two different types of datasets, sequences and expressions, for clustering genes and precisely predict functions of unknown genes using the clusters obtained. The effectiveness of our method was validated by its advantage over a lot of other methods in performance comparison.

The datasets for genes we used in our experiments were sequences and microarray expressions, both of which can be measured for any gene basically. We note that currently available biological data can be divided into roughly two classes: 1) relatively static data that include gene/protein sequences and phylogenetic profiles, and 2) biologically active (dynamic) data which are obtained by experimental observations like gene expressions and protein-protein interactions. A favorable aspect on the choice of gene sequences and microarray expressions is that we selected one dataset from each of the above two classes. In this sense, we can say that our method generated gene clusters by balancing between static and dynamic information on functional activities of genes.

We used the Spellman gene expression dataset [33] in our experiments. In out method, it is easy to replace the Spellman dataset with another dataset of gene expressions. If we choose another type of expression dataset, we will be able to include another type of dynamic information on genes in the combination of two data sources. Thus interesting future work would be to use some specific experimental conditions of gene expressions to annotate gene functions in the corresponding cell condition. We further note that our framework allows to use any type of existing functional categories of genes such as FunCat [26] and Pfam [10] as standard gene functions, instead of GO which we used in our experiments. Thus we can choose any of them if it is more reliable and covers a lot of genes.

A biological dataset can be categorized into either of structured or unstructured data. Thus our method is able to use any addition as well as the current two datasets, meaning that we can combine more than two different types of datasets, although the addition might be not necessarily applied to any gene.

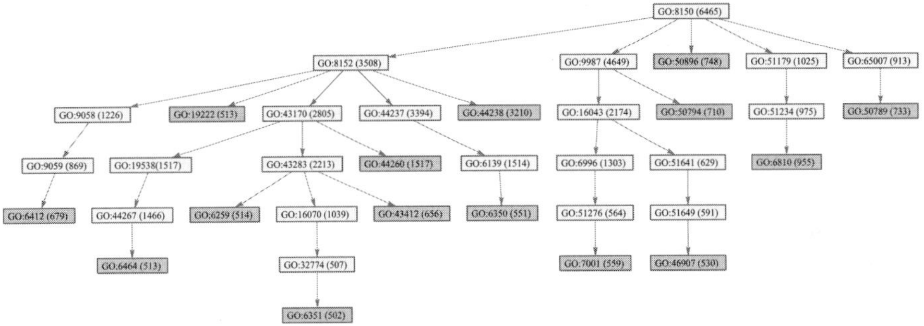

Fig. 6: Fifteen clusters obtained from GO.

In Section 3.5, we fix the size of clusters at ten, since in this setting the analysis by using FunCat on the clustering results was relatively easy. However, the GO information of ten clusters might be too vague to hypothesize the functions of unknown genes. If we use a larger number of clusters, we might be able to predict more detailed functions of unknown genes. We emphasize that in our framework, changing this parameter (K) is easy, and in fact we already performed the experiments when K is fifteen and twenty. Function prediction results under $K = 15$ and 20 are, of course, available upon request.

We used the data of *Saccharomyces Cerevisiae* only in our experiments. Obviously our annotation method is not specialized to this species but a general method, which can be applied to any species if we can make a gene network of sequence similarities from the sequence information of this species and have a microarray dataset on this species. Thus it would be also interesting to annotate genes of a more unfamiliar species with a lot of uncharacterized genes if for this species, there are datasets from which we can have enough information. From a different viewpoint, interesting future work might be to predict functions of genes of more than one species by using information on sequences and expressions of a wide variety of species at once.

Acknowledgements

The authors would like to thank Shanfeng Zhu, Ichigaku Takigawa and Raymond Wan for their helpful comments and fruitful discussion over this work. M. S. was supported in part by Kyoto University 21st Century COE Program "Knowledge Information Infrastructure for Genome Science", Grant-in-Aid for Young Scientists (B) by MEXT, Japan and BIRD of Japan Science and Technology Agency (JST). W-K. C. is supported in part by HKRGC Grant No. 7017/07P, HKUCRGC Grants, HKU Strategy Research Theme fund on Computational Sciences, Hung Hing Ying

Physical Research Sciences Research Grant, National Natural Science Foundation of China Grant No. 10971075 and Guangdong Provincial Natural Science Grant No. 9151063101000021. H. M. has been supported in part by BIRD of Japan Science and Technology Agency (JST).

Appendix A.   Generating Standard Fifteen Clusters from Gene Ontology (GO)

The Gene Ontology (GO) consists of controlled vocabularies describing three aspects of gene product functions: (i) molecular function, (ii) biological process and (iii) cellular component. Each of these three aspects is called an ontology and represented by a directed acyclic graph (DAG) where each GO term is attached to a node in the graph. The root (or the starting point) exists in a DAG, and GO terms are arranged hierarchically from general ones to specific ones in a DAG. This means that if a gene is assigned to one term (or a node), then the gene can be assigned to all ancestors of this term. Thus the number of genes assigned to a node is not less than that to any of its children.

To build a sequence network, we generate standard clusters using GO in the following manner: We first place a cut-off value, which is the minimal cluster size. We then start from the root to its descendants, and if the number of genes attached to one node is less than this cut-off value, we go back to its parent and stop there so that the number of genes assigned to this node is more than the cut-off value. By repeating this procedure, we have a set of clusters, keeping the minimum cluster size being larger than the specified cut-off value. We note that we can change the minimum cluster size of the cluster set by changing the cut-off value.

In our experiments, we used the cut-off value of 500 by which all genes in GO were divided into fifteen clusters. Figure 6 shows a tree with the fifteen clusters corresponding to the leaves (colored green) of this tree, and the GO term (accession ID) and the number of genes at each node are shown for all nodes in this tree. The ontology for each GO accession ID in Fig. 6 is described in Table A1.

References

[1] Azuaje, F., Clustering-based approaches to discovering and visualising microarray data patterns. *Briefings in Bioinformatics*, 4(1):31–42, 2003.

[2] Barabási, A-L., Reka., A., Emergence of scaling in random networks. *Science*, 286:509–512, 1999.

[3] Basu, S., Bilenko, M., Mooney, R. J., A probabilistic framework for semi-supervised clustering. *Proc. Tenth ACM SIGKDD*, 59–68, 2004.

[4] Boyle, E. I., *et al.*, Go::TermFinder–open source software for accessing gene ontology information and finding significantly enriched gene ontology terms associated with a list of genes. *Bioinformatics*, 20(18):3710–3715, 2004.

[5] Brem, R. B., *et al.*, Genetic dissection of transcriptional regulation in budding yeast. *Science*, 296:752–755, 2002.

[6] Brown, M., *et al.*, Knowledge-based analysis of microarray gene expression data by using support vector machines. *Proc. Natl. Acad. Sci.*, 97:262–267, 2000.

[7] Cai, C. Z., *et al.*, SVM-prot: web-based support vector machine software for functional classification of a protein from its primary sequence. *Nucl. Acids Res.*, 31(13):3692–3697, 2003.

[8] Chan, P. K., Schlag, M. D. F., Zien, J. Y., Spectral K-way ratio-cut partitioning and clustering. *IEEE Transactions on Computer-Aided Design of Integrated Circuits and Systems*, 13(9):1088–1096, 1994.

[9] Ding, C., *et al.*, A min-max cut algorithm for graph partitioning and data clustering. *Proc. IEEE International Conference on Data Mining*, 107–114, 2001.

[10] Finn, R. D., *et al.*, Pfam: clans, web tools and services. *Nucleic. Acids Res.*, 34(Database issue):D247–51, 2006.

[11] Gasch, A. P., *et al.*, Genomic expression programs in the response of yeast cells to environmental changes. *Mol. Biol. Cell*, 11(122):4241–4257, 2000.

[12] Guimera, R., Nunes Amaral, L. A., Functional cartography of complex metabolic networks. *Nature*, 433(7028):895–9, 2005.

[13] Guimera, R., Sales-Pardo, M., Amaral, L. A. N., Modularity from fluctuations in random graphs and complex networks. *Phys. Rev. E*, 70:025101, 2004.

[14] Guldener, U., *et al.*, CYGD: the comprehensive yeast genome database. *Nucl. Acids Res.*, 33(Database issue):D364–D368, 2005.

[15] Hagen, L., Kahng, A. B., New spectral methods for ratio cut partitioning and clustering. *IEEE Transactions on Computed Aided Design of Integrated Circuits and Systems*, 11:1074–1085, 1992.

[16] Hand, D. J., Till, R. J., A simple generalization of the area under the ROC curve for multiple class classification problems. *Machine Learning*, 45:171–186, 2001.

[17] Hastie, T., Tibshirani, R., Friedman, J. H., *The Elements of Statistical Learning.* Springer, 2003.

[18] Hughes, T. R., *et al.*, Functional discovery via a compendium of expression profiles. *Cell*, 102(1):109–126, 2000.

[19] Mamitsuka, H., Selecting features in microarray classification using ROC curves. *Pattern Recognition*, 39(12):2393–2404, 2006.

[20] Nash, R., *et al.*, Expanded protein information at sgd: new pages and proteome browser. *Nucl. Acids Res.*, 35(Database issue):D468–D471, 2007.

[21] Newman, M. E. J., Fast algorithm for detecting community structure in networks. *Physical Review E*, 69:066133, 2004.

[22] Newman, M. E. J., Girvan, M., Finding and evaluating community structure in networks. *Phys. Rev. E*, 69:026113, 2004.

[23] Ng, A. Y., Jordan, M. I., Weiss, Y., On spectral clustering: Analysis and an algorithm. *Advances in Neural Information Processing Systems 14*, 849–856. 2002.

[24] Ravasz, E., *et al.*, Hierarchical organization of modularity in metabolic networks. *Science*, 297(5589):1551–1555, 2002.

[25] Rost, B., Enzyme function less conserved than anticipated. *J. Mol. Biol.*, 318:595–608, 2000.

[26] Ruepp, A., The FunCat, a functional annotation scheme for systematic classification of proteins from whole genomes. *Nucl. Acids Res.*, 32:5539–5545, 2004.

[27] Sharan, R., Ulitsky, I., Shamir, R., Network-based prediction of protein function. *Molecular Systems Biology*, 3(88):1–13, 2007.

[28] Shi, J., Malik, J., Normalized cuts and image segmentation. *IEEE Trans. on Pattern Analysis and Machine Intelligence*, 22:888–905, 2000.

[29] Shiga, M., Takigawa, I, Mamitsuka, H., Annotating gene function by combining expression data with a modular gene network. *Bioinformatics*, 23(13):i459–i467, 2007.

[30] Shiga, M., Takigawa, I., Mamitsuka, H., A spectral clustering approach to optimally

combining numerical vectors with a modular network. *Proc. Thirteenth ACM SIGKDD*, 647–656, 2007.

[31] Smith T. F., Waterman, M. S., Identification of common molecular subsequences. *J. Mol. Biol.*, 147:195–197, 1981.

[32] Song, C., Havlin, S., Makse, H. A., Self-similarity of complex networks. *Nature*, 433:392–395, 2005.

[33] Spellman, P. T., *et al.*, Comprehensive identification of cell cycle-regulated genes of the yeast saccharomyces cerevisiae by microarray hybridization. *Mol. Biol. Cell*, 92(12):3273–3297, 1998.

[34] Storey, J. D., Akey, J. M., Kruglyak, L. Multiple locus linkage analysis of genomewide expression in yeast. *PLoS Biol.*, 3:e267, 2005.

[35] Troyanskaya, O., *et al.*, Missing value estimation methods for dna microarrays. *Bioinformatics*, 17(6):520–525, 2001.

[36] von Luxburg, U., A tutorial on spectral clustering. *Statistics and Computing*, 17(4):395–416, 2007.

[37] Wagstaff, K., Cardie, C., Clustering with instance-level constraints. *Proc. Seventeenth International Conference on Machine Learning*, 1103–1110, 2000.

[38] Wagstaff, K., *et al.*, Constrained k-means clustering with background knowledge. *Proc. Eighteenth International Conference on Machine Learning*, 577–584, 2001.

[39] Watts, D. J., Strogatz, S. H., Collective dynamics of 'small-world' networks. *Nature*, 393:440–442, 1998.

[40] White, S., Smyth, P., A spectral clustering approach to finding communities in graphs. *Proc. Fifth SIAM International Conference on Data Mining*, 76–84, 2005.

[41] Wu, L. F., *et al.*, Large-scale prediction of Saccharomyces cerevisiae gene function using overlapping transcriptional clusters. *Nature Genetics*, 31(3):255–265, 2002.

[42] Yvert, G., *et al.*, Transacting regulatory variation in saccharomyces cerevisiae and the role of transcription factors. *Nature Genetics*, 35(1):57–64, 2003.

[43] Zhong, S., Ghosh, J., A unified framework for model-based clustering. *Journal of Machine Learning Research*, 4:11–1037, 2003.

Table 6: Annotation for 151 uncharacterized genes in yeast genome

C	Uncharacterized genes	GO terms	**F**-value
C_1	YLR289w	GO:6412	0.8000
		GO:43283	0.7333
		GO:16043	0.5333
C_2	YBR028c YDL025c YGL083w YGR052w YJL057c YKL161c YKL168c YMR291w YPL141c YPL150w YPL236c YPR106w	GO:6468	0.7255
C_3	YDL048c YER130c YGL131c YGR067c YLR375w YML081w YPL230w YPR013c YPR015c YPR022c	GO:6357	0.5152
		GO:16043	0.5152
C_4	YBL066c YBR033w YBR150c YBR239c YDR520c YER184c YFL052w YFR057w YIL130w YJL016w YJL206c YKL222c YKR064w YLL054c YLR187w YLR278c YMR019w YNR063w YPR196w	GO:44237	0.5439
		GO:43283	0.5088
C_5	YER047c YFR023w YGR250c YJR061w YPL074w	GO:43170	0.6212
		GO:44238	0.6212
		GO:44237	0.5758
C_6	YFR038w YGR130c YGR196c YJR134c YKL050c YKL105c YKR016w YBR148w YCR051w YDL070w YDR026c YEL043w YER033c YFR016c YLR247c YMR031c YMR086w YPL009c YPL186c YPL2602 YPL263C	GO:6996	0.5225
		GO:44237	0.5225
		GO:44238	0.5112
C_7	YAR020c YBL111c YBL112c YBL113c YBR301w YCR104w YDR291w YDR332w YDR542w YEL049w YEL075c YEL076c YEL077c YER150w YER189w YFL020c YFL064c YFL065c YFL066c YGL261c YGR294w YHL046c YHL049c YHL050c YHR218w YHR219w YIL011w YIL176c YIL177c YJL223c YJL225c YKL224c YLL025w YLL064c YLL066c YLL067c YLR037c YLR461w YLR462w YLR464w YMR244w YNR076w YOL161c YOR394w YPL282w	GO:6139	0.6792
		GO:43283	0.6792
		GO:3996	0.6226
C_8	YBR108w YBR125c YFR022w YGR068c YIR003w YJL084c YKR021w YML118w YNL018c YNL034w YOL042w	GO:16043	0.6275
		GO:44238	0.6078
		GO:43283	0.5686
C_9	YAL065c YAR061w YAR062w YDL037c YDL211c YDR134c YFL051c YHR213w YHR214w YIL169c YJL078c YJL079c YJR151c YKR013w YMR317w YNL176c YOL036w YOR009w YOR227w	GO:16043	0.6216
C_{10}	YDR128w YDR267c YER066w YKL121w YMR102c YNL035c YPL183c YPL247c	GO:6996	0.6852
		GO:43283	0.6667
		GO:6139	0.5370

Table A1: Accession ID and Corresponding Ontology.

GO accession ID	Ontology
6139	nucleobase, nucleoside, nucleotide and nucleic acid metabolism
6259	DNA metabolism
6350	transcription
6412	protein biosynthesis
6464	protein modification
6351	transcription, DNA-dependent
6810	transport
6996	organelle organization and biogenesis
7001	chromosome organization and biogenesis (sensu Eukaryota)
8150	biological process
8152	metabolism
9058	biosynthesis
9059	macromolecule biosynthesis
9987	cellular process
16043	cell organization and biogenesis
16070	RNA metabolism
19222	regulation of metabolism
19538	protein metabolism
32774	RNA biosynthesis
43170	macromolecule metabolism
44237	cellular metabolism
43283	biopolymer metabolism
43412	biopolymer modification
44238	primary metabolism
44260	cellular macromolecule metabolism
44267	cellular protein metabolism
46907	intracellular transport
50784	cocaine catabolism
50794	regulation of cellular process
50896	response to stimulus
51179	localization
51234	establishment of localization
51276	chromosome organization and biogenesis
51641	cellular localization
51649	establishment of cellular localization
65007	biological regulation

A NOVEL META-ANALYSIS APPROACH OF CANCER TRANSCRIPTOMES REVEALS PREVAILING TRANSCRIPTIONAL NETWORKS IN CANCER CELLS

ATSUSHI NIIDA

aniida@ims.u-tokyo.ac.jp

SEIYA IMOTO

imoto@ims.u-tokyo.ac.jp

MASAO NAGASAKI

masao@ims.u-tokyo.ac.jp

RUI YAMAGUCHI

ruiy@ims.u-tokyo.ac.jp

SATORU MIYANO

miyano@ims.u-tokyo.ac.jp

Human Genome Center, Institute of Medical Science, University of Tokyo, 4-6-1 Shirokanedai, Minato-ku, Tokyo 108-8639, Japan

Although microarray technology has revealed transcriptomic diversities underlining various cancer phenotypes, transcriptional programs controlling them have not been well elucidated. To decode transcriptional programs governing cancer transcriptomes, we have recently developed a computational method termed EEM, which searches for expression modules from prescribed gene sets defined by prior biological knowledge like TF binding motifs. In this paper, we extend our EEM approach to predict cancer transcriptional networks. Starting from functional TF binding motifs and expression modules identified by EEM, we predict cancer transcriptional networks containing regulatory TFs, associated GO terms, and interactions between TF binding motifs. To systematically analyze transcriptional programs in broad types of cancer, we applied our EEM-based network prediction method to 122 microarray datasets collected from public databases. The data sets contain about 15000 experiments for tumor samples of various tissue origins including breast, colon, lung etc. This EEM based meta-analysis successfully revealed a prevailing cancer transcriptional network which functions in a large fraction of cancer transcriptomes; they include cell-cycle and immune related sub-networks. This study demonstrates broad applicability of EEM, and opens a way to comprehensive understanding of transcriptional networks in cancer cells.

Keywords: microarray data; cancer; expression module; transcriptional network.

1. Introduction

In the last decade, microarray technology has revealed transcriptomic diversities underlining various cancer phenotypes; on the other hand, we have yet little knowledge about transcriptional programs controlling them. In our previous study, we proposed a computational method termed EEM, which aims to identify expression modules. An expression module is defined as co-expressed genes under a common regulatory program (e.g., target genes of the same TF), and could be a key to understanding the regulatory program. EEM starts from prescribed gene sets defined by prior biological knowledge like TF binding motifs [10]; for each gene set, EEM first identifies the largest co-expressed subset of genes in an input microarray dataset. Using the size of the subset as a test statistic, EEM then obtains significant co-expressed subsets of genes as expression modules.

Recently, the amount of transcriptome data deposited in public databases is exploding. For example, GEO [15], which is one of the principle repositories for microarray data, returned 3024 data sets when queried for "cancer" (as of Aug. 2009). Although even only separate analysis of each dataset has yielded many significant findings, it is expected that meta-analysis of the massive datasets will provide more global insights into regulatory programs encoded in cancer transcriptomes.

In this study, we present a new version of the EEM method which enables efficient screening for expression modules, aiming at meta-analysis of a large number of cancer microarray data sets. Although the original version of EEM employs z-score calculation to evaluate coherence of each gene set, we newly introduce an efficient p-value calculation method based on the extreme value distribution. Moreover, we extend our EEM approach to systematic prediction of cancer transcriptional networks. To reveal prevailing cancer transcriptional networks, we applied this new version of EEM to 122 microarray datasets including about 15000 experiments for tumor samples of various tissue origins. This analysis successfully revealed a prevailing cancer transcriptional network which functions in a large fraction of cancer transcriptomes. Taken together, this study provides a foundation for network-level meta-analysis based on our EEM method.

2. Methods

2.1. *Overview of the EEM Algorithm*

First we briefly review the EEM algorithm, which is described in detail in Niida *et al.* [10]. EEM takes two types of input data: a microarray dataset and a gene set (we call a *seed gene set*). Let $E = \{e_1, \ldots, e_n\}$ be the set of gene expression profiles in the input microarray dataset, where $e_i = (e_{i1}, \ldots, e_{im})$ is a vector of the expression values of the i-th gene, i.e., e_{ij} is the expression value of the i-th gene in the j-th sample. We then assume that e_i exists as a point in a continuous m-dimensional gene expression space $\mathcal{S}$. EEM operates on a subset $E_M \subseteq E$, which is the expression profiles of the genes in the seed gene set. For a given radius parameter r and point $x \in \mathcal{S}$, define

$$C_x = \{e_i \in E_M : d(e_i, x) \leq r\}, \tag{1}$$

where d is the Euclidean distance. We call C_x a *coherent set*, and the point x the *center* of C_x. First, EEM finds the maximal sized coherent set C_B (and corresponding center B) for the genes in E_M. Next, by assuming the size of the maximal sized coherent set $|C_B|$ as a test statistics (we call the *EEM statistics*), EEM evaluates expression coherence of E_M. If it was judged to be significant, EEM obtains C_B as an expression module.

Using EEM, we can screen for expression modules in the transcriptome of the input microarray dataset. Prior to the screening, we prepare a set of gene sets that could contain expression module, based on prior biological knowledge like TF binding motifs. EEM then applies the above procedure to each of the gene sets, and finally obtains expression modules from gene sets of significant coherence.

2.2. *Calculation of P-Values*

In the original version, EEM calculated a Z score of an EEM statistic $|C_B|$ from an empirical null distribution of the statistics. The empirical null distribution can be generated by repeatedly calculating test statistics for randomly sampled gene sets whose size is equal to that of the input gene set. Based on the empirical null distribution, we can also calculate p-values; the empirical approach enables us to calculate p-values for any statistics without theoretical models for null distributions. However, if it relies only on the empirical null distribution, precise calculations of small p-values need prohibitive computational time for generation a large number of null statistics. Especially, speed of computation is essential in meta-analysis which deals with a large amount of transcriptome data.

To overcome this limitation, we propose a novel efficient p-value calculation method based on based on an extreme value distribution fitted to the empirical null distribution. Prior to the p-value calculation based on the extreme value distribution, we roughly calculate p-values to filter out non-significant genes. EEM calculates a p-value from an empirical null distribution from 100 randomly sampled genes sets. The seed gene sets with p-values > 0.05 are filtered out in this step. This step works well enough to filter out non-significant gene set before proceeding to the following computer intensive steps which calculate a larger number of null statistics.

Extreme value distributions are used to model the minimum or the maximum of the collection of random observations from the same arbitrary distribution [4]. Because an EEM test statistic is the size of the largest coherent subset among many possible candidates, it is reasonable to model the null distribution based on an extreme value distribution. A family of extreme value distributions is represented as the generalized extreme value distribution, which has the following cumulative distribution function:

$$F(x; \mu, \sigma. \xi) = \exp\left[-\left\{1 + \xi\left(\frac{x-\mu}{\sigma}\right)\right\}^{-1/\xi}\right]$$

for $1 + \xi(x - \mu)/\sigma > 0$, where $\mu \in \mathbf{R}, \sigma > 0$, and $\xi \in \mathbf{R}$ are the location, scale, and shape parameters, respectively.

We can fit the above equation to the empirical distribution, and use it to calculate p-values $(p = 1 - F(|C_B|))$. In practice, the empirical distribution is generated from 10^3 randomly sampled genes sets and the fitting was performed using R library *ismev*. As shown in Fig.1, we confirmed that p-values based on the fitted extreme value distribution correspond well to those based on the empirical distribution. Note that we can calculate p-value $< 10^{-3}$ by the fitted distribution, though it only requires the same computational time for sampling 10^3 null statistics. Combined with the first filtering step, the computational time is further boosted; for a microarray dataset of typical size in this study, EEM takes a few hours using a workstation with Intel Core i7 and 4G RAM.

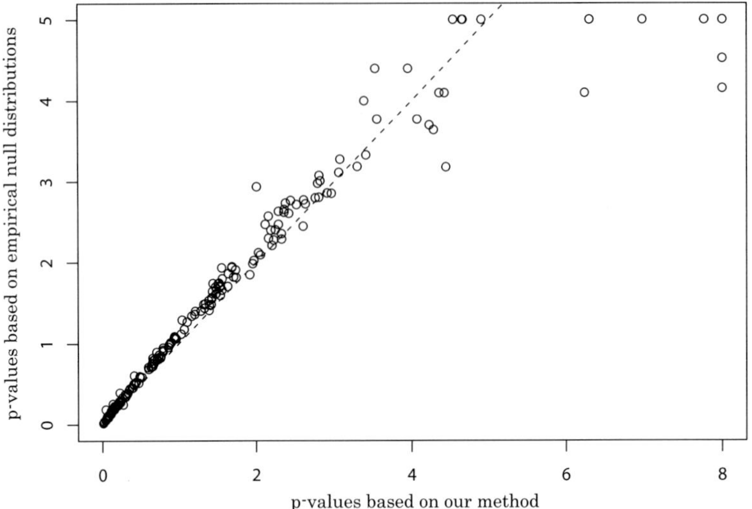

Fig. 1: Comparison of p-values based on fitted extreme distributions and empirical null distributions. we plot minus log p-values calculated by our method and 100000 random samplings of null EEM statistics. Although this result was obtained from the microarray dataset GSE3494 and TF binding motif-based gene sets, similar results were also obtained from different inputs.

2.3. *Input Data*

2.3.1. *Microarray Datasets*

In this study, we applied EEM to a large number of cancer microarray data sets. To obtain the cancer microarray data sets, we first collected the datasets from GEO [1] and ArrayExpress [11] using "cancer" as the keyword. The datasets were then excluded if the platform is not the the Affymetrix U133 series and the number of samples is less than 50. We finally obtained 122 datasets and applied the following preprocessing steps to them. Assuming the 5% lowest expression value as a floor, we rounded up expression values below the floor, and then converted expression values to the logarithmic scale. The Probe set IDs were converted to gene symbols. In cases that one gene symbol matches multiple probe set IDs, the probe set which shows the largest variance across the samples was mapped to the gene. A variation filter was then applied to the data, and we obtained the top half of genes with the highest variance. The expression profiles of them were normalized across samples and subjected to EEM analysis.

2.3.2. *Seed Gene Sets*

As the other type of input data, EEM needs seed gene sets that could contain coherent subset of genes. In this study, we prepared seed gene sets based on common TF binding motifs in their promoters [10]. Briefly, we prepared human and mouse promoter sequences encompassing the 500 bp upstream and 100 bp downstream of the transcription start sites. We also obtained 200 PWMs After motif clustering was applied to all vertebrate PWMs in TRANSFAC 2009.1 [8] to remove redundant motifs. For each PWM, we scored every human and mouse promoter sequence based on maximum log odds scores, and obtained the average of human and mouse homolog promoter scores as the PWM score for each gene. We assumed genes which record the 5 % highest PWM scores as seed gene sets sharing common TF binding motifs associated with the PWM.

2.4. *Transcriptional Network Prediction*

When given a microarray data set and seed gene sets based on TF binding motifs, EEM identifies not only expression modules, but also functional TF binding motifs regulating them. Starting from these EEM results, we can predict a transcriptional network in the transcriptome of the input microarray dataset. The predicted transcriptional network consists of three types of nodes: TF binding motifs (*motif nodes*), TFs (*TF nodes*), and GO terms (*GO nodes*). We assume that there are four types of edges between them.

(1) *MMI edges*: Undirected edges between motif nodes which represent motif-motif interactions
(2) *TF-binding edges*: Directed edges from TF nodes to motif nodes which represent TF biding to motifs
(3) *cis-regulatory edges*: Directed edges from motif nodes to TF or GO nodes which represents *cis*-transcriptional regulation
(4) *PPI edges*: Undirected edges between TF nodes which represent protein-protein interactions

Next, we explain how to derive these four types of edges from EEM results and various other data sources.

2.4.1. *MMI Edges*

MMI edges were predicted from overlaps between motif-associated expression modules. If two different motifs share a significantly large number of target genes, they should interact functionally, possibly, via physical interactions of binding TFs.

We tested overlaps of expression modules based on the hypergeometric distribution, and assumed that edges exists if p-values are smaller than 10^{-10}. Note that the TRANSFAC database includes redundant PWMs which can recognize the same binding motif. In such cases, expression modules derived from redundant PWMs can also have significant overlap with each other. Although we reduced the redundancy by clustering in the preparation step of PWMs, we found that the input PWM still includes some "cognate PWMs" which are annotated as bound to the same TF. Therefore, apparently significant MMIs might be "pseudo-MMIs" between a pair of redundant PWMs. To discriminate real MMIs from pseudo-MMIs, we measured similarity between each pair of motifs, and we assumed that the pair constitutes real MMI if the KL-distance exceeds 15 [10]. This cutoff value is based on the observation that it can filter out most of cognate PWM pairs.

2.4.2. *TF Binding Edges*

TF binding edges are predicted based on two information sources: TRANSFAC and expression profiles. Based on binding TF information attached to PWM entries in TRANSFAC, we first obtain binding TFs of each motif. Because the TRANSFAC database is based only on the primary literature and unsatisfactory from the viewpoint of comprehensiveness, we expanded TF binding information for each PWM as follows:

(1) If an associated binding TF is not human genes, we converted it to human homologs based on HomoloGene data.
(2) Because it is known that most members of the same TF family recognize similar binding motifs, we assumed that the other members which belong to the same family with a binding TF are also associated with the same motif. Gene family information was prepared from Ensembl.
(3) Note that input PWMs are representatives of clusters based on motif similarity. We also assigned TF associated with cluster members to the representatives.

To find binding TFs which functions in the transcriptome of the input microarray dataset, we measured correlation between the expression profile of each binding TF and the mean expression profile of module genes regulated by the motif, using sample label permutation tests; we assumed that a TF binding edges exists between the motif and a TF, if a p-value is less than 10^{-8}. Note that there are two types of TF binding edges; positive correlations are assumed as activating TF binding edges while negative correlations are assumed as repressive TF binding edges.

2.4.3. *Cis-Regulatory Edges*

Cis-regulatory edges can be obtained from results from EEM analysis. If an expression module associated with a motif includes a TF, the motif was assumed to regulate expression of the TF. GO terms enriched motif-associated expression modules were connected to motifs (p-value $< 10^{-6}$).

2.4.4. *PPI Edges*

PPI edges are based on PPI data registered by public databases. The data were prepared as described in Tamada *et al.* [14]

2.4.5. *Construction of Transcriptional Networks*

Starting from motif nodes with p-values $< 10^{-6}$ in EEM analysis, we construct a transcriptional network for an input microarray dataset. These "seed" motif nodes is connected to each other by MMI edges, and to TF nodes by TF-binding edge. We depicted *cis*-regulatory edges between the seed motifs and the added TF nodes, and PPI edges between the added TFs. GO nodes are added to the seed motifs via *cis*-regulatory edges. The network predicted in this way can be assumed as a transcriptional network which functions in the cancer transcriptome of the input microarray dataset.

Moreover, to reveal a functional transcriptional networks across multiple microarray datasets, we can obtain a "meta-network" by superimposing networks predicted for each microarray dataset. We counted the number of nodes and edges across networks for all the microarray datasets, and constructed a network by assembling nodes and edges which are predicted more frequently than a prespecified cutoff (3 out of 122 in this study). Network visualizations are performed so that the node size and the edge width proportionally reflect frequency of their prediction.

3. Results and Discussion

EEM analysis identifies expression modules that could be keys to understanding transcriptional programs in the transcriptome of the input microarray dataset. We predicted transcriptional networks, starting from expression modules identified by EEM and TF binding motif associated with them; we constructs a transcriptional network by connecting motifs based on module overlap, and adding regulatory TFs to the motifs based on the TRANSFAC database and expression correlations. PPI and GO terms enriched in each module were also incorporated to predicted networks. Furthermore, we attempted network-level meta-analysis to reveal prevailing cancer transcriptional networks, which possibly regulate fundamental oncogenic processes commonly employed by various types of cancers. We downloaded 122 microarray datasets associated the "cancer" keyword from GEO and ArrayExpress, and predicted transcriptional networks for each dataset. We then depicted a "meta-network" by assembling nodes and edges which were repeatedly predicted across the microarray datasets.

Fig. 2 shows the meta-network, where node size and edge width reflect frequency of their prediction. The meta-network contains two major sub-networks. One sub-network (lower right) includes E2F and DP family TFs, NFYA, and their binding motifs. It has been known, and also confirmed by our GO enrichment analysis, that the E2F and DP families are master regulators of cell-cycle [9]; it is very reasonable that they drive one of the principal transcriptional programs in cancer transcriptomes. This sub-network harbors positive auto-regulatory loops mediated by several E2F family genes, suggesting that they act as an engine in cell-cycle regulation. The other sub-network (upper left) contains various TF and motif nodes in addition to two main nodes, IRF and PU.1 binding motifs; they apparently act as hubs in this network and might be drug targets in cancer therapy. This sub-network includes known cancer genes like IRF, ETS, and RUNX [3, 6, 13] and is linked to immune-system by GO analysis. Although this result suggests that this immune transcriptional program is responsible for oncogenesis in broad types of cancer cells, it is also possible that a part of this network is due to infiltrating immune cells in the tumor microenvironment.

Our EEM-based network prediction takes as input various types of fragmented knowledge deposited in the databases (e.g. TF binding motifs, binding TF information associated with motifs, GO, PPI, etc.) and systematically assembles them into networks which presumably functions in the transcriptome of an input microarray dataset. Although many other types of network prediction methods have been proposed for transcriptome analysis, our approach possesses outstanding features as compared to the others. We can predict direct casual relationships based on TF binding information, which cannot be predicted by co-expression or Bayesian network approaches which rely on only transcriptome data [2, 5, 7]. Our predicted network also includes motif-motif interaction, PPI, GO terms associated with motifs; they make the predicted networks information-rich, highly interpretable and easy to extract biological knowledge. In this study, we applied EEM-based network prediction to more than a hundred cancer microarray datasets and, by superimposing networks predicted for each dataset, constructed a "meta-network". This meta-network approach should increase reliability of predicted networks; in fact, the obtained meta-network includes many TF associated with cancer, and are highly consistent with the accumulated knowledge of molecular biology. Although a number of studies have performed meta-analysis of cancer transcriptomes [12, 16], they have mainly focused on individual genes or a set of genes associated with oncogenic phenotypes. In contrast, our meta-analysis focused for the first time on cancer transcriptional networks, and successfully revealed prevailing transcriptional network in cancer cells.

Our analysis successfully identified two major transcriptional programs employed by various types of cancer cells: cell-cycle and immune-related transcriptional programs. While our previous study also demonstrated that the breast cancer transcriptome harbors similar cell-cycle and immune-related transcriptional programs, this study expanded the finding to many other types of cancer, suggesting their generality in various types of cancer. Based on our approach developed in this study, our next challenge should be to find transcriptional programs specific to each types of cancer. To address this problem, we need an enough number of datasets for each type of cancer; we are currently collecting microarray datasets more comprehensively for future studies.

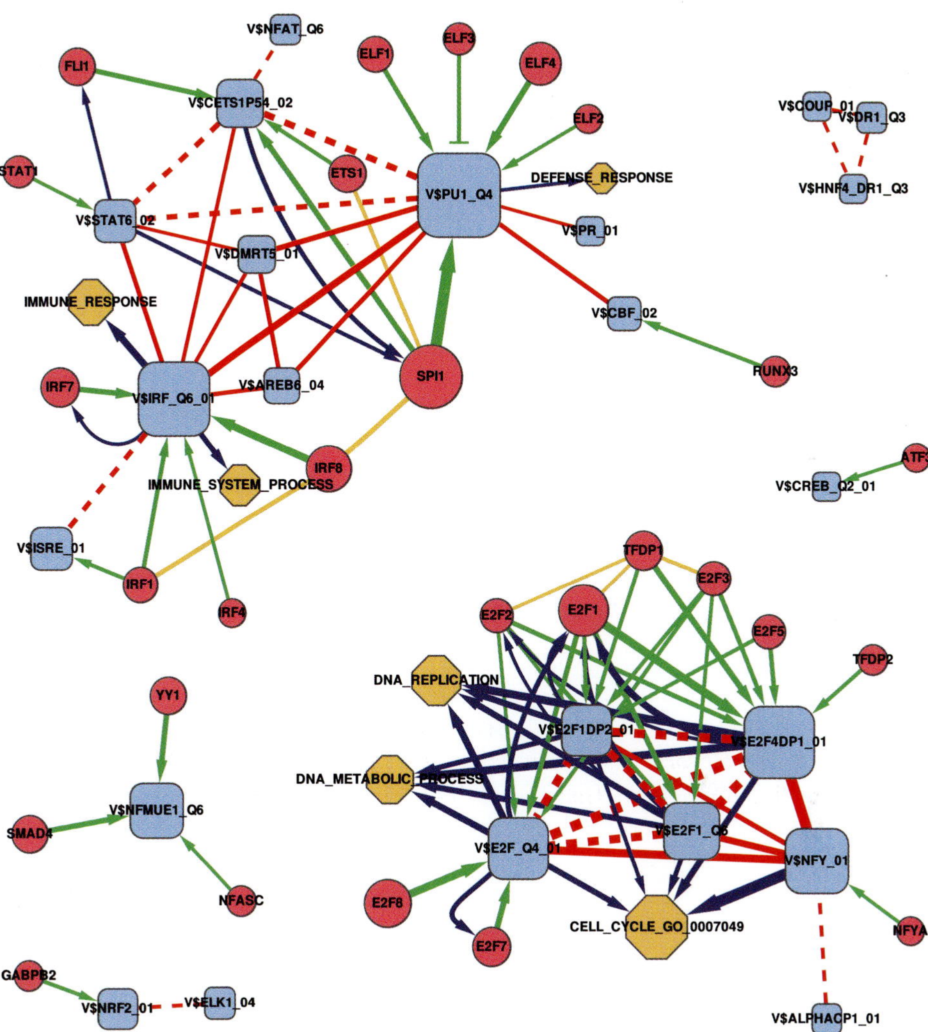

Fig. 2: Prevailing transcriptional networks in cancer cells. EEM based network prediction were applied to 122 cancer microarray datasets and frequently predicted nodes and edges were displayed as a "meta-network". Node size and edge width proportionally reflect frequency of their prediction. Pink, light blue, and yellow nodes denote TF, motif, and GO nodes, respectively. Solid red edges indicate MMI edges while dashed red edges indicate "pseudo-MMI edges" which might be bound by the same TF. Green normal directed edges indicate activating TF binding edges while green directed edges terminated with small bars indicate repressive TF binding edges. Blue and yellow edges represent *cis*-regulatory and PPI edges, respectively.

To pursue the comprehensiveness of analysis, we need to expand not only microarray datasets, but also seed gene sets. In this study, seed gene sets are based on only TF binding motifs on the proximal promoter regions. Clearly, this point is a major drawback of this study; we cannot obtain a real global view of oncogenic transcriptional programs without considering distal *cis*-regulation. In previous study, we demonstrated that seed gene sets prepared from ChIP-chip data can be also input to EEM analysis; one of solutions of this problem is preparing ChIP-chip-derived seed gene sets by literature curation. As new technologies recently start to reveal a genomic view of *cis*-regulatory regions [15], another possible solution is expected; more global prediction of transcriptional targets will be feasible in the near future by combining TF binding motifs with comprehensive information about of *cis*-regulatory regions.

4. Conclusion

In this study, we presented a novel meta-analysis approach of cancer transcriptome data. Application of an extended version of EEM to more than a hundred microarray datasets revealed prevailing transcriptional networks which functions in various types of cancer. This study is the first meta-analysis to focus on cancer transcriptional networks, and has opened a way to comprehensive understanding of transcriptional networks in cancer cells.

References

[1] Barrett, T., Troup, D., Wilhite, S., Ledoux, P., Rudnev, D., Evangelista, C., Kim, I., Soboleva, A., Tomashevsky, M., Marshall, K., Phillippy, K., Sherman, P., Muertter, R., Edgar, R. Ncbi geo: archive for high-throughput functional genomic data. *Nucleic Acids Res*, 37:D885–90, 2009.

[2] Basso, K., Margolin, A., Stolovitzky, G., Klein, U., Dalla-Favera, R., Califano, A. Reverse engineering of regulatory networks in human b cells. *Nat Genet*, 37:382–90, 2005.

[3] Blyth, K., Cameron, E., Neil, J. The runx genes: gain or loss of function in cancer. *Nat Rev Cancer*, 5:376–87, 2005.

[4] Coles, S. An introduction to statistical modeling of extreme values. *London: Springer-Verlag*, 2001.

[5] Friedman, N., Linial, M., Nachman, I., Pe'er, D. Using bayesian networks to analyze expression data. *J Comput Biol*, 7:601–20, 2000.

[6] Honda, K., Taniguchi, T. Irfs: master regulators of signalling by toll-like receptors and cytosolic pattern-recognition receptors. *Nat Rev Immunol*, 6:644–58, 2006.

[7] Lee, H., Hsu, A., Sajdak, J., Qin, J., Pavlidis, P. Coexpression analysis of human genes across many microarray data sets. *Genome Res*, 14:1085–94, 2004.

[8] Matys, V., Kel-Margoulis, O., Fricke, E., Liebich, I., Land, S., Barre-Dirrie, A., Reuter, I., Chekmenev, D., Krull, M., Hornischer, K., Voss, N., Stegmaier, P., Lewicki-Potapov, B., Saxel, H., Kel, A., Wingender, E. Transfac and its module transcompel: transcriptional gene regulation in eukaryotes. *Nucleic Acids Res*, 34:D108–10, 2006.

[9] Nevins, J. The rb/e2f pathway and cancer. *Hum Mol Genet*, 10:699–703, 2001.

[10] Niida, A., Smith, A., Imoto, S., Aburatani, H., Zhang, M., Akiyama, T. Gene set-based module discovery in the breast cancer transcriptome. *BMC Bioinformatics*, 10:71, 2009.

[11] Parkinson, H., Kapushesky, M., Kolesnikov, N., Rustici, G., Shojatalab, M., Abeygunawardena, N., Berube, H., Dylag, M., Emam, I., Farne, A., Holloway, E., Lukk, M., Malone, J., Mani, R., Pilicheva, E., Rayner, T., Rezwan, F., Sharma, A., Williams, E., Bradley, X., Adamusiak, T., Brandizi, M., Burdett, T., Coulson, R., Krestyaninova, M., Kurnosov, P., Maguire, E., Neogi, S., Rocca-Serra, P., Sansone, S., Sklyar, N., Zhao, M., Sarkans, U., Brazma, A. Arrayexpress update–from an archive of functional genomics experiments to the atlas of gene expression. *Nucleic Acids Res*, 37:D868–72, 2009.

[12] Rhodes, D., Yu, J., Shanker, K., Deshpande, N., Varambally, R., Ghosh, D., Barrette, T., Pandey, A., Chinnaiyan, A. Large-scale meta-analysis of cancer microarray data identifies common transcriptional profiles of neoplastic transformation and progression. *Proc Natl Acad Sci U S A*, 101:9309–14, 2004.

[13] Sharrocks, A. The ets-domain transcription factor family. *Nat Rev Mol Cell Biol*, 2:827–37, 2001.

[14] Tamada, Y., Araki, H., Imoto, S., Nagasaki, M., Doi, A., Nakanishi, Y., Tomiyasu, Y., Yasuda, K., Dunmore, B., Sanders, D., Humphreys, S., Print, C., Charnock-Jones, D., Tashiro, K., Kuhara, S., Miyano, S. Unraveling dynamic activities of autocrine pathways that control drug-response transcriptome networks. *Pac Symp Biocomput*, 14:251–63, 2009.

[15] Visel, A., Blow, M., Li, Z., Zhang, T., Akiyama, J., Holt, A., Plajzer-Frick, I., Shoukry, M., Wright, C., Chen, F., Afzal, V., Ren, B., Rubin, E., Pennacchio, L. Chip-seq accurately predicts tissue-specific activity of enhancers. *Nature*, 12;457:854–8, 2009.

[16] Xu, L., Geman, D., Winslow, R. Large-scale integration of cancer microarray data identifies a robust common cancer signature. *BMC Bioinformatics*, 8:275, 2007.

NEW KERNEL METHODS FOR PHENOTYPE PREDICTION FROM GENOTYPE DATA

RITSUKO ONUKI[1]   TETSUO SHIBUYA[2]
onuki@hgc.jp   tshibuya@hgc.jp

MINORU KANEHISA[1,2]
kanehisa@kuicr.kyoto-u.ac.jp

[1] *Bioinformatics Center, Institute for Chemical Research, Kyoto University, Gokosho, Uji, Kyoto 611-0011, Japan*
[2] *Human Genome Center, Institute of Medical Science, University of Tokyo, 4-6-1 Shirokanedai, Minato-ku, Tokyo 108-8639, Japan*

Phenotype prediction from genotype data is one of the most important issues in computational genetics. In this work, we propose a new kernel (i.e., an SVM: Support Vector Machine) method for phenotype prediction from genotype data. In our method, we first infer multiple suboptimal haplotype candidates from each genotype by using the HMM (Hidden Markov Model), and the kernel matrix is computed based on the predicted haplotype candidates and their emission probabilities from the HMM. We validated the performance of our method through experiments on several datasets: One is an artificially constructed dataset via a program GeneArtisan, others are a real dataset of the NAT2 gene from the international HapMap project, and a real dataset of genotypes of diseased individuals. The experiments show that our method is superior to ordinary naive kernel methods (i.e., not based on haplotype prediction), especially in cases of strong LD (linkage disequilibrium) .

Keywords: genotypes; SVM; phasing; single-nucleotide polymorphism.

1. Introduction

Most variations of the DNA sequences in humans are at single-base sites, in which more than one nucleic acid can be observed across the population [5]. These sites (i.e. loci) are called SNPs (single nucleotide polymorphisms). The sequence of pairs of DNA residues at each SNP site is called genotypes. The DNA residues which consist of genotypes are called alleles. There are two kinds of allele, major allele and minor allele. The major allele is an allele which is observed more frequently than another allele at the SNP site. The minor allele is less frequently observed than major allele at the SNP site. A sequence of contiguous alleles of SNPs on the same chromosome is called a haplotype. Most experimental techniques for determining SNPs do not provide haplotype information [4], [8]. The experiments generate only an unordered pair of allele readings for each site on the two chromosomes, i.e., a genotype. To obtain the haplotype data, we need to infer the haplotypes from genotype data. The process of inferring haplotypes from genotypes is called phasing or resolving.

For the tailor-made medicine, the variations of the individuals' phenotypes caused by the differences of DNA sequences become important and it is becoming important to predict individuals' phenotypes from genotypes. In this work, we developed new kernel methods to predict phenotype from genotype data, utilizing phased haplotype information. Then, we applied our methods to the disease status prediction and the prediction of the individual's geographic background. Previous work for SVM-based disease status prediction is based on the number of the minor allele for each SNP site [2]. We focused on the phased haplotype and discussed whether the phased haplotype can improve the prediction accuracies.

There are many software tools for haplotype inference: PHASE [20], [21], GERBIL [13], HAP [9] and HIT [19]. Among the tools listed above, HIT can provide not only the optimal haplotype but multiple candidates of haplotypes. Based on the multiple candidates of haplotypes obtained by HIT, we propose new kernel methods.

In the last part of the paper, we demonstrate the effectiveness of these methods through experiments on several datasets, i.e., two kinds of real genotype datasets and an artificial dataset.

2. Methods

2.1. *Preliminaries*

In this section, we describe the notations and 2 methods, i.e., Haplotype Inference Technique (HIT) [19] and haplotype similarity measure, which are the basis of our methods.

Notations

Consider n genotypes over m SNP loci from the same chromosome. These loci are numbered $1, \cdots, m$ from left to right in the physical order. In most cases, only two alternative bases (i.e. alleles) occur at a SNP site. The allele for the SNP locus is an element of the set $A = \{1,0\}$ where 1 and 0 refer to the most frequent allele and others at each SNP locus respectively. Sometimes the data contains missing values, and we represent them by '?' . Then a haplotype is a sequence in A^m (e.g. 1010111100). A genotype is a sequence of unphased (i.e., unordered) allele pairs and is defined as a sequence in A'^m, where $A' = A \times A$ (e.g. 00 00 00 11 01 11 0? 00 00 11). Thus a genotype data consists of $2n \times m$ values. Given a set of genotypes, the phasing problem is to find their corresponding most probable haplotype pairs (i.e. diplotypes) that could have generated the genotypes.

Haplotype Inference Technique (HIT)

HIT is an algorithm that uses HMM (Hidden Markov Model) [17], [18] to estimate haplotypes from genotypes. It shows relatively high accuracy among other software programs [19]. From now on, we describe how HIT infers haplotypes from genotypes. In HIT, we assume that all sites are bi-allelic.

The model parameters, i.e., transition probabilities and emission probabilities, are estimated by the EM algorithm, using only the input genotype dataset. This means the

HIT does not need any training dataset. The number of founders K is a parameter we need to set before the HMM is trained. In Rastas et al. [19], they say that the HIT gives good results if K is set to any value larger than 4. They themselves use the setting K = 7 in their experiments against the Daly et al.'s data [6]. So we also set K to 7 in our experiments in section 3. Once the model has been trained, we can estimate haplotypes from genotypes. Moreover we can obtain multiple haplotype candidates with emission probabilities that the HMM emits the haplotype.

Haplotype similarity measure

We use the hamming distance as a similarity measure between haplotypes. For a haplotype $h \in A^m$, let $h(k)$ denote the allele at locus k of the haplotype h. The similarity measure between two haplotypes h and h' is defined as:

$$s\left(h, h'\right) = \sum_{k=1}^{m} I\left(h(k), h'(k)\right) \tag{1}$$

where $I(a, b) = 0$ if alleles a and b are the same and $I(a, b) = 1$ otherwise. Tzeng et al. [22] and Li and Jiang [14] also use this hamming distance or its variant as the similarity measure between haplotypes.

To obtain length-independent measure, we consider the following value $d(h, h')$ as the similarity between haplotypes h and h':

$$d(h, h') = \frac{s(h, h')}{m}. \tag{2}$$

This measure may not be always best for all evolutionary or practical scenarios. It should be noted that our methods described in section 2.2 can use any other similarity measures between haplotypes, though there are not known any standard measures other than the measure described above.

2.2. *Our method*

In this section, we first introduce genotype-genotype distance defined in our another work [15] for computing kernels between genotypes (Section 2.2.1). Based on the kernels, we next describe how we predict phenotypes from genotype data in Section 2.2.2.

2.2.1. *Genotype distance*

To develop new kernels, we need somewhat distance between genotypes. We introduce genotype-genotype distance, haplotype frequency-based distance (HFD), proposed in [15].

Before defining the genotype-genotype distance, we define a distance between haplotype pairs based on the haplotype distance described in Section 2.1. Let $a = (h_1, h_2)$ and $a' = (h_1', h_2')$ be two haplotype pairs to be compared, where $h_1, h_2, h'_1, h'_2 \in A^m$. We define the distance between haplotype pairs a and a' as:

$$H(a, a') = \min\{\, d(h_1, h_1') + d(h_2, h_2'), \qquad d(h_1, h_2') + d(h_2, h_1')\}. \qquad (3)$$

Using this distance, we propose the following genotype-genotype distance.

Haplotype frequency-based distance (HFD)

For genotypes $g, g' \in A'^m$, let $c_i = (h_{i1}, h_{i2})$ $(1 \leq i \leq M)$ and $c_j' = \left(h_{j1}', h_{j2}'\right)$ $(1 \leq j \leq M')$ be candidate haplotype pairs for g and g' respectively, which are computed by HIT. Note that these candidate sets are not all the set of possible candidates, as there are usually an exponential number of candidates when we infer haplotypes from genotypes. The M and M' are user-specified upperbounds of the numbers of candidates to be enumerated by the HIT. In the experiments in Section 4, we let M = M' = 100.

Let p_i, p_j' be the emission probabilities of the candidate haplotype pairs, c_i and c_j'. The emission probabilities are considered as the haplotype frequencies. The haplotype frequency-based distance (HFD) between genotypes is defined as the following summation:

$$HFD(g, g') = \sum_{j=1}^{M'} \sum_{i=1}^{M} H\left(c_i, c_j'\right) \cdot q_i \cdot q_j' \qquad (4)$$

where $q_i = p_i / \sum_{k=1}^{M} p_k$ and $q_j' = p_j' / \sum_{k=1}^{M'} p_k'$. q_i and q_j' are the normalized emission probabilities of the candidate haplotype pairs, c_i and c_j', respectively.

2.2.2. *Our learning algorithm*

We propose a learning algorithm for phenotype prediction from genotypes which consists of 3 steps. The learning algorithm needs a training dataset and a test dataset. We train our learning algorithm with the training dataset and test the leaning algorithm with the test dataset. The details of each step are described as follows.

Step1. Haplotype Inference

For each of the given training set and test set of genotypes, we extract candidates of haplotype pairs and their corresponding emission probabilities by using HIT. In the experiments in section 3, we let the number of the founders for the HMM be 7, which is the same as the setting in the experiments in Rastas et al. [19].

Step2. Computation of the kernels

Using the result of the step 1, we evaluate HFD against all the pairs of genotypes of the training set and the test set. The details of HFD are described in the previous section 2.2.1. Based on the HFD, we compute the kernels in the 2 ways. One is $e^{-HFD(g, g')}$ for genotypes g and g', which we call exponential HFD and another is $1 - HFD(g, g')$ for genotypes g and g′, which we call linear HFD. In most cases, our kernels are positive-semi definite. If the kernels are not the positive-semi definite, we do as the follows.

Let K be the kernel matrix. K can be written as $K = PMP^{-1}$, where M is a triangular matrix and its diagonal elements are the eigenvalues of K, and P is an orthogonal matrix. We replace the negative diagonal elements of M by zero and let it be M′. Let $K' = PM'P^{-1}$. We use this K′ as the kernel matrix.

Step3. Phenotype Prediction by SVM

We predict the phenotypes from the test set of genotype data with SVM [24] based on the kernels computed in the step 2 of our learning algorithm.

3.   Data sets

3.1.   *The NAT2 dataset*

In the next section, we do experiments against a set of NAT2 gene-related genotypes [3] taken from the HapMap datasets of the version on March 1, 2007 [11]. The data consists of 270 genotypes with 24 SNP sites. The 270 genotypes can be divided into 3 ethnic groups (i.e. populations). The first group consists of genotypes of 90 Utah residents with ancestry from northern and western Europe (CEU), the second group consists of genotypes of 45 unrelated individuals of Han Chinese in Beijing and 45 unrelated individuals of Tokyo, Japan (CHB+JPT), and the last group consists of genotypes of 90 Yoruba people in Ibadan Nigeria, West Africa (YRI).

The NAT2 gene are said to be related to the susceptibility to some toxicities and cancers [7], [10]. Thus it is very important to study the differences of the NAT 2 genes among different populations to elucidate the ethnic difference in such susceptibilities.

3.2. *Disease datasets*

Dataset of Crohn's disease [6] is derived from the 616 kb region of human Chromosome 5q31 that may contain a genetic variant responsible for Crohn's disease by genotyping 103 SNPs for 129 trios. All offspring belong to the case population, while almost all parents belong to the control population. In entire data, there are 144 case and 243 control individuals.

Dataset of autoimmune disorder [23] is sequenced from 330kb of human DNA containing gene CD28, CTLA4 and ICOS. These genes are proved related to autoimmune disorder. A total of 108 SNPs were genotyped in 384 cases of autoimmune disorder and 652 controls.

3.3. *Simulated dataset*

We also applied our algorithms to an artificial genotype dataset to examine the performance of our methods. The data is generated by GeneArtisan [25] based on the

Coalescent model. The region includes one SNP site related to a genetic risk factor for diseases.

We made training datasets of 1000 genotypes, where 500 are affected (i.e. case) genotypes and 500 are normal (i.e. control) genotypes, using the above tool.

4. Results and Discussions

In the following sections, we show the results of the 3-fold cross-validations. Throughout the prediction validations, we compare our methods with other methods, i.e. Brinza' method [2], SVM-fisher [12] and the method based on the number of the major allele. SVM-Fisher is a HMM based method. We used the same HMM in the step 1 of our algorithm for SVM-Fisher. The method based on the number of the major allele counts the number of the major allele at each SNP site and makes the feature vector of each individual's genotype.

4.1. *Results on high LD datasets*

Our HFD based methods show high accuracy on YRI and CEU datasets from HapMap datasets (Table 1, Table 2). We found that the datasets which our methods show high accuracies have high Linkage Disequilibrium (LD) (Figure 1). The LD measure a co-segregation between the SNP sites. It is estimated that the haplotypes of high LD datasets can be inferred more precisely than low LD datasets (i.e. CHB+JPT dataset) and our HFD based methods show high accuracies than the other methods. Our methods also show high accuracy on the simulated datasets. The simulated datasets were generated by the tool incorporating that the human genome consists of haplotype blocks [16]. The regions in the haplotype blocks show high LD. It is estimated that the simulated datasets tend to show the haplotype structure more clearly and the haplotypes can be inferred precisely than real datasets and accuracy of our HFD based method are the higher than any other methods. And the accuracy is the higher value than the other real datasets (Table 3).

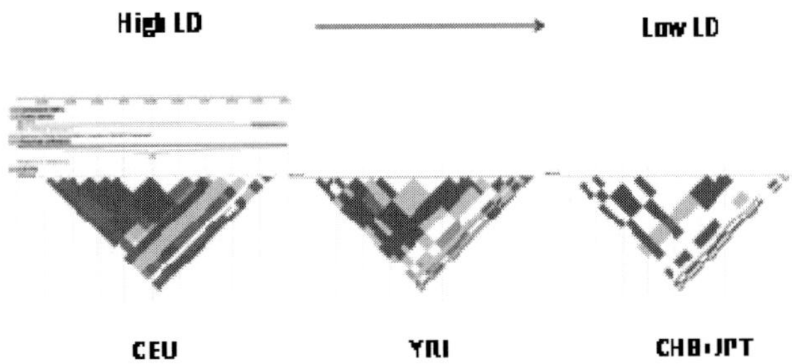

Figure 1. LD plots of NAT2 gene from HapMap by Haploview [1]. The black color and gray color show the high LD score, and the white color shows the lower LD score.

Table 1. Results of 3-fold cross-validation for CEU dataset.

Method	Sensitivity	Specificity	Accuracy
Exp HFD	0.822	0.828	0.832
Linear HFD	0.822	0.811	0.815
Major allele	0.811	0.800	0.804
SVM-Fisher	0.678	0.728	0.711
Brinza's method	0.733	0.755	0.748

Table 2. Results of 3-fold cross-validation for YRI dataset .

Method	Sensitivity	Specificity	Accuracy
Exp HFD	0.722	0.939	0.867
Linear HFD	0.722	0.939	0.867
Major allele	0.733	0.911	0.863
SVM-Fisher	0.489	0.861	0.737
Brinza's method	0.667	0.856	0.793

Table 3. Results of 3-fold cross-validation for simulated dataset.

Method	Sensitivity	Specificity	Accuracy
Linear HFD	0.960	1.000	0.980
Major allele	0.970	0.920	0.945
SVM-Fisher	0.980	0.620	0.800

4.2. *Results on low LD datasets*

As described in section 4.1, CHB+JPT dataset from HapMap shows lower LD than the other HapMap datasets, which we estimated is the reason why our HFD based methods don't show higher accuracies than the other methods (Table 4). Our HFD based methods also don't show high accuracies on the disease datasets (Table 5, Table 6). The gene related to the disease datasets are ranged in longer regions. The datasets are truncated by three haplotype blocks and the regions contain much lower LD [23], which we estimated is the reason why our HFD based methods don't show high accuracies.

Table 4. Results of 3-fold cross-validation for CHB+JPT dataset.

Method	Sensitivity	Specificity	Accuracy
Exp HFD	0.911	0.800	0.832
Linear HFD	0.911	0.783	0.826
Major allele	0.922	0.744	0.804
SVM-Fisher	0.767	0.728	0.778
Brinza's method	0.911	0.800	0.837

Table 5. Results of 3-fold cross-validation for Crohn' s disease dataset.

Method	Sensitivity	Specificity	Accuracy
Exp HFD	0.468	0.603	0.543
Linear HFD	0.561	0.536	0.545
Major allele	0.574	0.456	0.496
SVM-Fisher	0.668	0.352	0.447
Brinza's method	0.388	0.575	0.501

Table 6. Results of 3-fold cross-validation for dataset of Autoimmune disorder.

Method	Sensitivity	Specificity	Accuracy
Exp HFD	0.530	0.521	0.525
Linear HFD	0.451	0.692	0.603
Major allele	0.425	0.684	0.587
SVM-Fisher	0.493	0.524	0.515
Brinza's method	0.601	0.570	0.580

5. Conclusion

We proposed kernel methods based on the support vector machine. We succeeded in accurate phenotype prediction especially for the high LD datasets, i.e. CEU and YRI dataset of NAT2 gene and simulated dataset. On the other hand, our results were not so good for the low LD datasets, i.e. JPT+CHB dataset of NAT2 gene and disease datasets. As our kernels are based on phased haplotypes, we concluded that our prediction accuracies depend on the degree of the LD of the datasets. We can infer the haplotype more precisely for the high LD datasets than low LD datasets in the step 1 in our methods. The prediction accuracy of haplotype inference influences the prediction accuracy of the SVM.

In this work, we only treated the disease status prediction. We will extend our models to take into account whether the individuals have the potential of the disease or not. We will also try to take into account the variations of the haplotype blocks to consider more biological aspects into our models.

Acknowledgments

We thank Dr. Ryo Yamada for comments on various problems in genetics related to this research. We also thank Dr. Yoshihiro Yamanishi and Dr. Hisashi Kashima for very helpful discussions.

References

[1] Barrett, J.C., Fry, B., Maller, J., Daly, M.J., Haploview: analysis and visualization of LD and haplotype maps, *Bioinformatics,* 21:263-265, 2004.

[2] Brinza, D., Zelikovsky, A., Discrete Methods for Association Search and Status Prediction in Genotyep Case-Control Studies, *Proc. of IEEE 7-th International symposium on Bioinformatics and Bioengineering* 270-277, 2007

[3] Butcher, N.J., Boulouvala, S., Sim, E. and Minchin, R.F., Pharmacogenetics of the arylamine N-acetyltransferases, *Pharmacogenomics,* 2:30-42, 2002.

[4] Clark, V.J., Metheny, N., Dean, M., Peterson, R.J., Statistical Estimation and pedigree analysis of CCR2-CCR5 haplotypes, *Hum Genet,* 108:484-496, 2001.

[5] Collins, F.S., et al., A DNA polymorphism discovery resource for reserch on human genetic variation, *Genome Res.,* 8:1229-1231, 1998.

[6] Daly, M., Rioux, J., Hudson, T., Lander, E., High-resolution haplotype structure in human genome, *Nat Genet,* 29:229-232, 2001.

[7] Evans, D.A., N-acetyltransferase, *Pharmacol Ther* 42:157-234, 1989.

[8] Fallin, D., Schork, N.J., Accuracy of haplotype frequency estimation for biallelic loci, via expectation-maximization algorithm for unphased diploid genotype data, *Am J Hum Genet,* 67:947-959, 2000.

[9] Halperin, E., Eskin, E., Haplotype reconstruction from genotype data using imperfect phylogeny, *Bioinformatics,* 20: 104-113, 2004.

[10] Ito, T., Inoue, E., Kamatani, N., Association Test Algorithm Between a Qualitative Phenotype and a Haplotype or Haplotype Set Using Simultaneous Estimation of Haplotype Frequencies, Diplotype Configurations and Diplotype-Based Penetrances, *Genetics,* 168: 2339-2348, 2004.

[11] International HapMap Consortium, The International Hapmap Project, *Nature,* 426:789-796, 2003. http//www.hapmap.org

[12] Jakkola, T., Diekhans, M., Haussler, D., A Discriminative Framework for Detecting Remote Protein Homologies, *Journal of computational biology,* 7:95-114, 2000.

[13] Kimmel, G., Shamir, R., GERBIL: Genotype resolution and block identification using likelihood, *Proc Nat Acad Sci,* 102;158-162, 1998.

[14] Li, J., Jiang, T., Haplotype-based linkage disequilibrium mapping via direct data mining, *Bioinformatics,* 21(24):4384-4393, 2005.

[15] Onuki R., Shibuya, T., Yamada, R., Kanehisa, M., A new measure of inter-diplotype distance using haplotype frequency in populations, in preparation.

[16] Phillip, M.S., Lawrence, R., Sachidanandam, R., Morris, A.P., Balding, D.J., Donaldson, M.A., Studebaker, J.F., et al. Chromosome-wide distribution of haplotype blocks and the role of recombination hot spots, *Nat Genet,* 33:382-387, 2003.

[17] Rabiner, L.R., Juang, B.H., An Introduction to Hidden Markov Models, *IEEE ASSP Mag.,* 3(1):4-16, 1986.

[18] Rabiner, L.R., A tutorial on hidden Markov models and selected applications in speech recognition, *Proceedings of the IEEE,* 77: 257-285, 1989.

[19] Rastas, P., Koivisto, M., Mannila, H., Ukkonen, E., A hidden markov technique for haplotype reconstruction, *Lecture Notes in Bioinformatics,* 3692:140-151, 2005.

[20] Stephans, M., Smith, N., Donnelly, P., 2001. A new statistical method for haplotype reconstruction from population data, *Am J of Hum Genet,* 68:978-989, 2001.

[21] Stephans, M., Scheet, P., Accounting for decay of linkage disequilibrium in haplotype inference and missing-data imputation, *Am J of Hum Genet,* 76:449-462, 2005.

[22] Tzeng, J.Y., Devlin, B., Wasserman, L., Roeder, K., On the identification of disease mutations by the analysis of haplotype similarity and goodness of fit, *Am J of Hum Genet,* 72:891-902, 2003.

[23] Ueda, H., Howson, J.M.M., Esposito, L., et al. 2003. Association of the T cell Regulatory Gene CTLA4 with Susceptibility to Autoimmune Disease, *Nature,* 423 506-511, 2003.

[24] Vapnik, V., 1998. Statistical Learning Theory, *Wiley, NY,* 1998.

[25] Wang, Y., Rannala, B., *In Silico* Analysis of Disease-Association mapping Strategies Using the Coalescent Process and Incorporating Ascertainment and Selection, *Am J of Hum Genet,* 76:1066-1073, 2005.

EFFICIENT AND DETAILED MODEL OF THE LOCAL Ca^{2+} RELEASE UNIT IN THE VENTRICULAR CARDIAC MYOCYTE

THOMAS SCHENDEL

thomas.schendel@web.de

MARTIN FALCKE

martin.falcke@mdc-berlin.de

Max Delbrück Centre of Moleculare Medicine, Robert-Rössle-Str. 10, Berlin, Germany, 13092

We present here an efficient but detailed approach to modelling Ca^{2+}-induced Ca^{2+} release in the diadic cleft of cardiac ventricular myocytes. In this framework we developed a spatial resolved Ca^{2+} release unit (CaRU), consisting of the junctional sarcoplasmic reticulum and the diadic cleft, with a well defined channel placement. By taking advantage of time scale seperation, the model could be finally reduced to only one ordinary differential equation for describing Ca^{2+} fluxes and diffusion. Additionally the channel gating is described in a stochastic way. The resulting model is able to reproduce experimental findings like the gradedness of SR release, the voltage dependence of ECC gain and typical spark life time. Due to the numerical efficiency of the model, it is suitable to use for whole cell simulations. The approach we want to use extend the developed CaRU to such a whole cell model is already outlined in this work.

Keywords: excitation-contraction coupling; calcium dynamics; modelling.

1. Introduction

The key step of excitation-contraction coupling (ECC) in cardiac myocytes, namely the Ca^{2+}-induced Ca^{2+} release(CICR) is a local phenomen, which takes place in very small subcompartments, the so called diadic clefts. The diadic cleft is the space between the membrane of the T-tubules on the one side and the junctional sarcoplasmic reticulum (JSR) on the other side. Membrane depolarisation leads to an activation of the L-type Ca^{2+}-channels (LCCs) in the membrane of the T-tubules, which causes a Ca^{2+}-influx into the diadic cleft. This Ca^{2+} activates the ryanodine receptors (RyRs) at the sarcoplasmic reticulum (SR), which release Ca^{2+} from the SR. That causes a large concentration rise in the diadic cleft and a partial depletion of the JSR. The small geometric properties of the diadic cleft provides the strong coupling between membrane depolarisation and Ca^{2+}-release.

The Ca^{2+} dynamics in the whole myocyte is modulated by further mechanisms. The main points are here the removal of Ca^{2+} from the myoplasm into the extracellular space via the Na^{+}-Ca^{2+}-exchanger or the Ca^{2+}-pumps, the reloading of the sarcoplasmic reticulum (SR) via the SERCA pumps and finally the Ca^{2+} buffers inside the cell.

Modelling the Ca dynamics in the whole ventricular myocyte is a challenging task due to the different time and length scales of the dynamics. On the one side the local mechanism of CICR in the diadic clefts (the volume of each cleft is in the range of 10^{-18}l), on the other side the global dynamics (implementing the different compartments, fluxes between this compartments and Ca^{2+}-buffers) has to be described. Finally, the diadic clefts (5,000-20,000 per cell [3, 4, 18] has to be included into the global model. Due to this (numerically) sophisticated task, former whole cell models have only simple models of the diadic cleft, which sometimes neglect the stochastic nature of the channel gating of LCCs or RyRs [9, 23] or treat the diadic cleft as one compartment with a uniform Ca^{2+} concentration and do not consider any spatial detail [9, 14, 23]. One exception is reported in Ref. [6], where the diadic cleft was divided into 4 different spatial parts. But this model is compUtionally expensive.

Nevertheless, there are also very detailed models of only the diadic cleft. These models are usually spatially resolved and stochastic [11, 18] and some even take explicitly the protein structure into account [18]. Nevertheless, these very detailed models lack of any global dynamics and they are computationally too demanding to be used as a CaRU in a global model.

The aim of this paper is to close the gap between a detailed, but numerical expensive description of the diadic cleft on the one side and numerically effecient whole cell models with a simple description of the local Ca^{2+} release unit (CaRU) on the other side. Therefore we want to present a detailed model of the local CaRU that is numerically efficient enough to use it for whole cell models. In our model the diadic cleft is spatially extended, with a defined channel placement. The channel gating is modelled in a stochastic manner. We also take into account the Ca^{2+} dynamics in the JSR.

Our model is able to reproduce basic experimental observations like gradedness (different amount of Ca^{2+} influx will trigger a different Ca^{2+} release from the SR), decreasing gain (defined by the ratio of Ca^{2+} release from SR to Ca^{2+} influx through LCCs) by increasing the membrane voltage and typical spark life time. It is therefore suitable to be used as a local CaRU in whole cell simulations. The description of our model will already gives some insight into how we want to extent the model to simulate the whole ventricular myocyte in future.

2. Model Description

The model of the local CaRU consists of three compartments, the diadic cleft, JSR and the cytosol bulk (see Fig. 1A). The RyRs at the JSR are colocalized to the LCC in the membrane of the T-tubules. We assumed the diadic cleft to be a cylinder with a height of 15 nm and a radius of 100 nm [17]. Because of the comparably small height, we effectively reduced the problem to a two-dimensional one by neglecting gradients orthogonal to the membranes. For a mathematical description of the CaRU, we had to deal with three different dynamics: the dynamics of the channel states (of the LCCs and RyRs), the Ca^{2+} distribution in the diadic cleft and finally the Ca^{2+} dynamics in the JSR.

2.1. *Channel Placement and Gating*

We use 16 RyRs and 4 LCCs. The channel placement is seen in Fig. 1B. 4 RyRs are always close to one LCC, which reflects the experimental observation that the opening of one LCC can trigger the opening of 4-6 RyRs [20].

For the LCC we used the kinetic scheme from [9], which is a lumped state model. It consists of 3 states, open, closed and inactivated (Fig. 2A). Inactivation is in this scheme Ca^{2+} dependent. The rates have the following form [9]:

$$k_{CO} = \frac{e^{\frac{V-V_L}{dV_L}}}{t_L \left(e^{\frac{V-V_L}{dV_L}} + 1 \right)}$$

$$k_{OC} = \frac{\phi_L}{t_L}$$

$$k_{CI} = c_{di} \frac{e^{\frac{V-V_L}{dV_L}} + a}{K_L \tau_L \left(e^{\frac{V-V_L}{dV_L}} + 1 \right)}$$

$$k_{IC} = b \frac{e^{\frac{V-V_L}{dV_L}} + a}{\tau_L \left(e^{\frac{V-V_L}{dV_L}} + 1 \right)}$$

V...membrane voltage, c_{di}...Ca^{2+} concentration in the diadic cleft

For the RyRs we used a scheme developed in Ref. [2] (Fig. 2B). It consist of four states (rest, open and two inactivated states). The step to inactivation is Ca^{2+} dependent and the opening rate of the channel is a nonlinear function of Ca^{2+}.

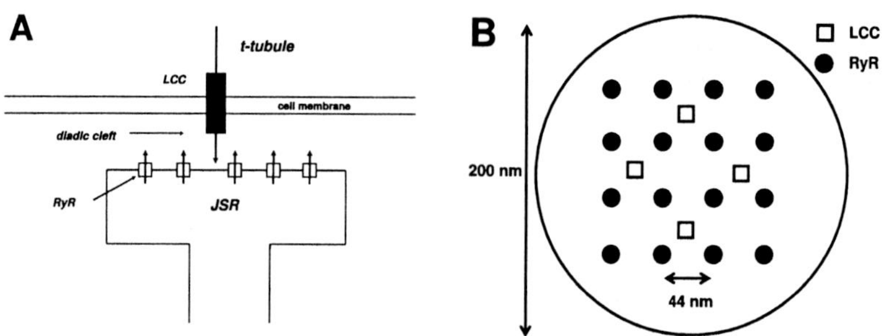

Fig. 1: (A) Scheme of the local CaRU, consisting of the diadic cleft and the JSR. RyRs and LCCs are colocalized. (B) The channel placement of 16 RyRs and 4 LCCs. Close to each LCC are 4 RyRs.

We decided to model the channel gating in a stochastic manner due to the experimental fact that a Ca^{2+} release event (Ca^{2+} spark) is built up by 1-7 open RyRs [21]. More details about the specific numerical treatment is provided in subsection 2D.

2.2. *Diffusion Profile in the Diadic Cleft*

Inside the diadic cleft we take into account the Ca^{2+} fluxes from LCCs and RyRs and Ca^{2+} diffusion ($c_{di}(\vec{r})$ denotes the Ca^{2+} concentration at the point $\vec{r}$ in the diadic cleft):

$$\frac{dc_{di}(\vec{r})}{dt} = \sum_i J_{LCC}^i(\vec{r_i}) + \sum_j J_{RyR}^j(\vec{r_j}) + D\Delta c_{di}(\vec{r}) \tag{1}$$

Because the typical diffusion time Δt for R=100 (radius of diadic cleft) nm is about $R = \sqrt{D\Delta t} \rightarrow \Delta t = 100\mu s$, which is much smaller than the typical time for the channel gating (usually in the range of ms), we can assume quasistationarity (see also [8]):

$$\frac{dc_{di}(\vec{r})}{dt} \approx 0 = \sum_i J_{LCC}^i(\vec{r_i}) + \sum_j J_{RyR}^j(\vec{r_j}) + D\Delta c_{di}(\vec{r}) \tag{2}$$

$$J_{LCC}^i = J_L \delta V \frac{c_{ext}e^{-\delta V} - c_{di}^i}{1 - e^{-\delta V}} \tag{3}$$

$$J_{RyR}^i = g(c_{JSR}^i - c_{di}^i) \tag{4}$$

$$c_{di}(r = R) = c_{bulk} \tag{5}$$

c_{JSR} denotes the Ca^{2+} concentration in the JSR, c_{ext} denotes the external Ca^{2+} concentration (set constant to $1000\mu M$). To simplify the following consideration, we will assume that c_{bulk} (Ca^{2+} concentration at the border of the diadic cleft) is constant and we will come back to this point at the end of this section. To calculate

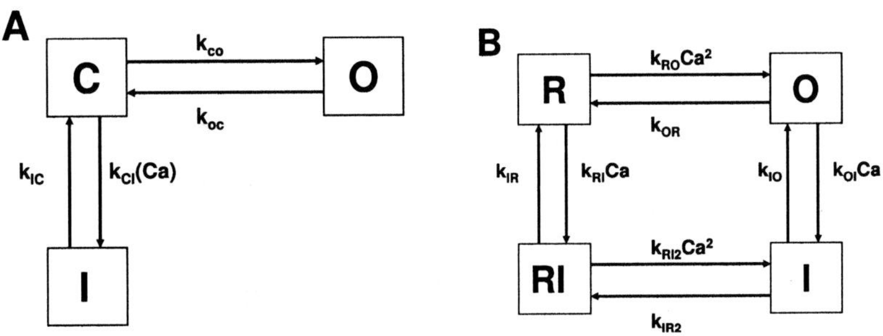

Fig. 2: Kinetic schemes for LCCs and RyRs. O denotes the open, C the closed, I the inactivated and R the resting state. (A) Kinetic scheme of the LCC from [9]. This scheme is a simplification of a more complex model. (B) Kinetic scheme from [2]. Activation is a nonlinear function of Ca^{2+}, Inactivation is Ca^{2+} dependent.

now the Ca^{2+} diffusion profile in the diadic cleft, we have to take the feedback of the concentration to the current into account: an opening of one channel will cause a flux which will lead to a Ca^{2+} concentration profile — which also affects the flux through this channel (as seen in Eq.(3)and (4)). In order to solve this problem, we obtained first an analytical solution for a simpler problem: one channel located in the middle of a circle with a radius r_c of $r_c = 6nm$ (typical radius of RyR) (we want to mention that the used fluxes and permeabilities are already scaled by the height h of the diadic cleft ($J = J_{real}/h$) in order to adjust to the two-dimensional problem):

$$-D\Delta c = \tilde{g}(c_{JSR} - c) \; ; \; r \leq r_c$$
$$-D\Delta c = 0 \; ; \; r > r_c \tag{6}$$

and for the boundary condition

$$c(R) = c_{bulk} \tag{7}$$

The solution reads($k = \sqrt{\frac{\tilde{g}}{D}}$):

$$c(r) = c_{JSR} - \frac{c_{JSR} - c_{bulk}}{I_1\left(k\,r_c\right) ln\left(\frac{R}{r_c}\right) k\,r_c + I_0\left(k\,r_c\right)} I_0\left(k\,r\right); \; r \leq r_c \tag{8}$$

$I_0(kr)$ denotes the modified Bessel function of *0th* order. With this result one can calculate the mean Ca^{2+} concentration $\bar{c}$ at the channel and that allows to obtain the flux J through the channel. We now replaced the spatially extended channel by a δ-source with the same flux J. In general we obtain for the source coordinate $r = r_i$:

$$c(r) - c_{bulk} = \frac{J}{\pi D} \sum_{k=0}^{\infty} \frac{B_0\left(x_{0,k}\frac{r_i}{R}\right) B_0\left(x_{0,k}\frac{r}{R}\right)}{x_{0,k}^2 J_1^2\left(x_{0,k}\right)}$$
$$+ \frac{2J}{\pi D} \sum_{n=1}^{\infty}\sum_{k=0}^{\infty} \frac{B_n\left(x_{n,k}\frac{r_i}{R}\right) B_n\left(x_{n,k}\frac{r}{R}\right) cos(n\varphi)}{x_{n,k}^2 J_{n+1}^2\left(x_{n,k}\right)}$$
$$= J\kappa(r_i, r) \tag{9}$$

B_n denotes the Bessel function of nth order. $x(n,k)$ are the kth roots of the besssel function of nth order.

For comparison with Eq.(8) we chose $r_i = 0$. To circumvent the problem of the divergent δ-function at $r = 0$ we had to calculate the distance Δr, which satiesfies $c(\Delta r) = \bar{c}$ for $r_i = 0$. With this result we were able to calculate approximately the currents for arbitrarily chosen r_i:

1. Calculate the angle $\Delta\varphi = \arccos\left(1 - (\Delta r)^2/2r_i^2\right)$ where the mean calcium concentration of the channel is measured ($\bar{c} \approx c(r_i, \Delta\varphi)$)

2. Calculate $\bar{c}$ for a given flux J. Define $\kappa_i := \kappa(\vec{r_i}, \vec{r_i} + \Delta\vec{r_i})$. κ_i can be determined

by using Eq.(9): $\bar{c} = c_{bulk} + \kappa_i J$.

3. $J = g(c_{JSR} - \bar{c}) = g(c_{JSR} - c_{bulk} - \kappa_i J)$ (in case of a RyR) $\rightarrow J = \frac{g(c_{JSR} - c_{bulk})}{1 + \kappa_i g} \rightarrow$
$\bar{c} = c_{bulk} + \left(\kappa_i g (c_{JSR} - c_{bulk}) \right) / \left(1 + \kappa_i g \right)$

The only approximation here is the use of the value of $|\Delta r|$ calculated with a channel in the centre of the cleft. If more than one channel is open, their produced Ca^{2+} concentration will affect each other. This situation ends up to a system of linear equations. For details, see Appendix 5.1.

2.3. *Ca^{2+} Dynamics in the JSR*

We assume that the JSR has a spatially uniform Ca^{2+} concentration. This Ca^{2+} concentration is affected by the outward flux J_{RyR} through the RyR and the refill flux from the network sarcoplasmic reticulum (NSR). We set the timescale for refilling to 30 ms [1]. Moreover, the Ca^{2+} buffer Calsequestrin was considered and for simplicity we chose the fast buffer approximation. We obtained:

$$\frac{dc_{JSR}}{dt} = \beta_{JSR} \left(-\frac{1}{V_{JSR}} \sum_j J^j_{RyR} + \frac{c_{NSR} - c_{JSR}}{\tau_{refill}} \right) \tag{10}$$

$$\beta_{JSR} = \left(1 + \frac{n K_c B_{CSQN}}{(K_c + c_{JSR})^2} \right)^{-1} \tag{11}$$

n...number of Ca^{2+}-binding sites of Calsequestrin, B_{CSQN}...concentration of Calsequestrin, K_C...dissociation constant of Calsequestrin (chosen from [14])
During all simulations, we keep the Ca^{2+} concentration in the NSR c_{NSR} constant ($c_{NSR} = 700 \mu M$). To speed up simulations, we linearized β_{JSR} for each time step Δt around the actual $\tilde{c}_{JSR}$, and obtained the following analytic expression:

$$c_{JSR} = \frac{\sqrt{-D}}{2a} \tanh \left(\operatorname{artanh} \left(\frac{2a\tilde{c}_{JSR} + b}{\sqrt{-D}} \right) - \frac{\sqrt{-D}\Delta t}{2} \right) - \frac{b}{2a} \tag{12}$$

where a, b, c and D are defined by:

$$\frac{dc_{JSR}}{dt} = ac_{JSR}^2 + bc_{JSR} + c$$

$$D = 4ac - b^2$$

2.4. *Matching the Dynamics Together*

On the one side we used a deterministic description of diffusion and the amount of flux through the open channels and on the other side we used a stochastic approach for the channel gating. Such a "mixed approach" was justified in [8]. In case of comparably quick channel state transitions, the Ca^{2+} concentration in the JSR (and therefore the Ca^{2+} profile in the diadic cleft) remains nearly unchanged between transitions and we could use the efficient Gillespie algorithm [5]. In case of slower channel transition the Ca^{2+} concentration changes in the JSR (and therefore the

Ca^{2+} concentration changes in the diadic cleft) are not negligible and we have to use a hybrid algorithm for the stochastic stimulation described in [15]. Our algoritm works as follows:

Assume that N channel transitions are possible with the associated propensities (rates) $\alpha_1...\alpha_N$. Following the Gillespie algorithm (assuming that the propensities are not changing) the time step τ until the next channel transition takes place is defined by:

$$\sum_i^N \alpha_i \tau = \ln 1/r_1 \tag{13}$$

r_1 is a uniform random number in the interval $[0,1]$. If during this time τ the Ca^{2+} concentration in the JSR remains nearly constant, we can apply the Gillespie algorithm. Otherwise the Ca^{2+} concentration in the diadic cleft and therefore the propensities α_i are time dependent. In this case the time τ is defined by ([15]):

$$\int_t^{t+\tau} \sum_i^N \alpha_i(s)ds = \ln 1/r_1 \tag{14}$$

In the simulations we divided the integral into smaller time steps during which the concentration in the JSR remains nearly constant.

2.5. *Calculating the Ca^{2+} Concentration at the Boundary of the Diadic Cleft*

The bulk Ca^{2+} concentration is mainly influenced by the diadic cleft itself. We assumed one diadic cleft (modelled as a point source) in the middle of the cell (modelled as a cylinder with a radius R_m of $R_m = 11\mu m$ and a length L of $L = 140\mu m$). We take into account the pumping of Ca^{2+} from the myoplasma into the NSR via SERCA, the mobile Ca^{2+} buffer Calmodulin (denoted in the equations as M for the unbounded form), the immobile Ca^{2+} buffer Troponin (denoted in the equations as T) and a constant leak flux from the NSR (that maintains in equilibrium a resting Ca^{2+} concentration of $0.1\mu M$):

$$\frac{\partial c}{\partial t} = D_c\Delta c + J_{leak} - g_{SERCA}\frac{c^2}{K_m^2 + c^2} - k_+^M cM + k_-^M(B_M - M)$$
$$- k_+^T cT + k_-^T(B_T - T)$$
$$\frac{\partial M}{\partial t} = D_M\Delta M - k_+^M cM + k_-^M(B_M - M)$$
$$\frac{\partial T}{\partial t} = -k_+^T cT + k_-^T(B_T - T) \tag{15}$$

We want to solve this system of PDEs by using three component Green's functions. Therefore we had to linearize the system of Eq.(15) around $c_{rest} = 0.1\mu M$ and we got:

$$\frac{\partial \delta c}{\partial t} = D_c \Delta \delta c - \rho_{SERCA} \delta c - k_+^M \delta c M_0 - k_+^M \delta M c_0 - k_+^T \delta c T_0 - k_+^T \delta T c_0$$

$$\frac{\partial \delta M}{\partial t} = D_M \Delta \delta M - k_+^M \delta c M_0 - k_+^M \delta M c_0$$

$$\frac{\partial \delta T}{\partial t} = -k_+^T \delta c T_0 - k_+^T \delta T c_0 \tag{16}$$

The Green's functions can be seen in the Appendix 5.2. By simulations we obtained that the stationary solution is sufficient at the boundary of the cleft to calculate the bulk Ca^{2+} concentration. That is not really surprising, because of the used small length scales where the diffusion is dominant (typical diffusion time for $R=100$ nm is 100 μs calculated in 2.2). The stationary cytosolic concentration at the boundary of the celft c_{bulk} is again proportional to the total flux J out of the diadic cleft: $c_{bulk} - c_{rest} = \beta J$ (the way to calculate β is shown in the Appendix 5.2). c_{bulk} enters the current and concentration calculation according to Eq. 17 in Appendix 5.1.

3. Results

The local CaRU has for it's validation to reproduce general experimental results. One important property is gradedness, which means that a different Ca^{2+} influx through LCCs will trigger a different amount of Ca^{2+} release from the SR. In Fig. 3A the integrated Ca^{2+} fluxes (over $100ms$) through RyRs and LCCs is depicted. It is seen that the model is able to exhibit gradedness. The maximum of Ca^{2+} release from SR is shifted by about $10mV$ with respect to the the maximum of Ca^{2+} entry through LCCs. This is due to the higher sensitivity of RyRs to higher single LCC currents (which occur after the Goldmann-Hodgkin-Katz equation (Eq.(3))

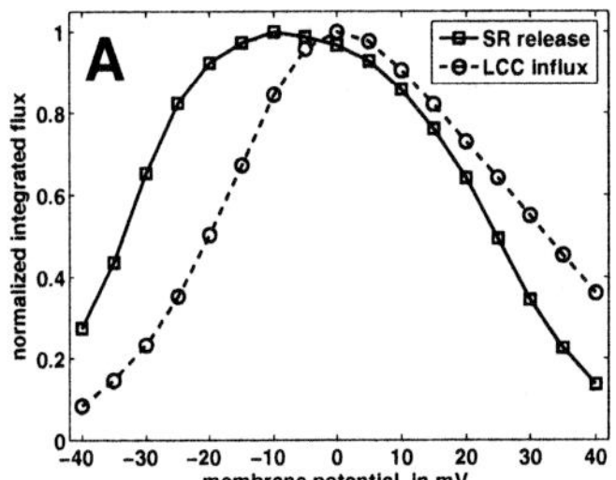
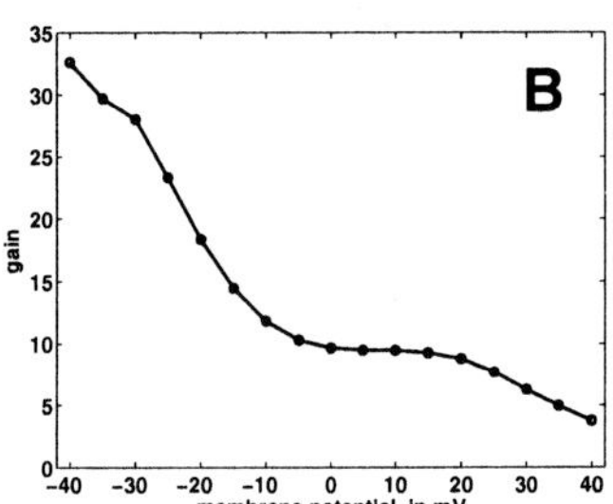

Fig. 3: A) Over 100 ms integrated Ca^{2+} fluxes through RyR and LCC as a function of membrane potential. The peak for Ca^{2+} release from SR is around 10 mV shifted from the peak of Ca^{2+} entry through LCC. (B) Gain defined as the ratio of Ca^{2+} release from SR to Ca^{2+} entry via LCC. It decreases for increasing membrane potential. Here are two figures side-by-side. (a) Figure caption for figure 2a. (b) Figure caption for figure 2b.

for smaller membrane potentials) and reproduces the experimental data [22]. If we now define the property gain as the ratio between the Ca^{2+} release from the SR to the Ca^{2+} entry from the LCC, we can plot it as a function of membrane potential. This curve is a decreasing function of membrane potential (Fig. 3B), which is in accordance with experiment [22].

We also studied the dependence of ECC for variable distances from the t-tubule to JSR. In Fig. 4A gain was plotted for three different heights. It is seen that the gain decreases for increasing height. Since gain is simply an indicator for the strength of the coupling between membrane depolarisation and Ca^{2+} release, ECC decreases with increasing height of the diadic cleft.

Finally, we studied the spark time distribution (Fig. 4B). We defined the time of a spark as the time between when the first RyR opens until all RyRs are closed. The mean spark life time was determined to $9.8ms$, which is in the range of experimental measurements ($11.6ms$ [21]).

4. Discussion

In this paper we have described a detailed model of the local CaRU that is numerically fast enough to be used for whole cell simulations. The model describes individual RyRs and LCCs as stochastic Markov chains, the concentration profiles inside the cleft, the interaction of channel currents resulting from them and the concentration dynamics in the JSR. Computational efficiency is achieved by a quasistatic approximation for the concentration profiles inside the cleft. The cleft model reproduces basic experimental findings like gain and gradedness, average spark life times and the dependence of gain on cleft height.

Former papers focusing on whole cell simulations often used only simple forms of the local CaRU, mostly without a spatially resolved diadic cleft. Hinch et al. [9] and Williams et al. [23] treat the diadic cleft only as a uniform single compartment, using deterministic models for LCC and RyR. In the work of Restrepo et al. [14]

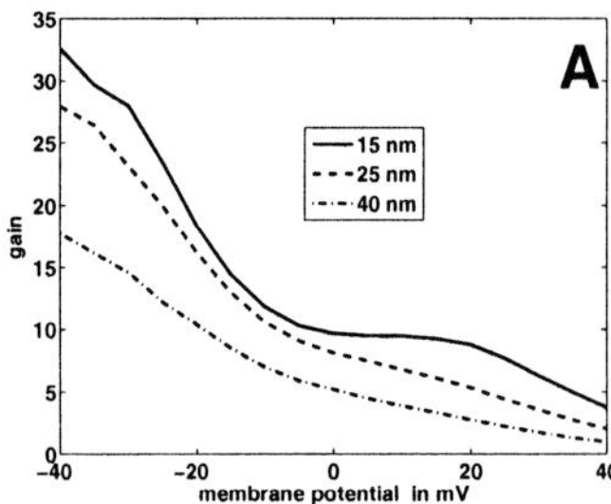

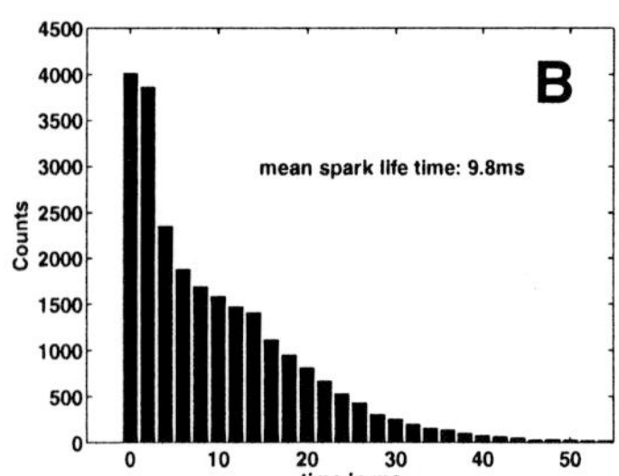

Fig. 4: (A) Gain for different heights of the diadic cleft: h=15, 25, 40 nm. The coupling decreases for increasing height. (B) Spark time distribution over an samplesize of 24261 sparks for V=0 mV and 15 nm height of the diadic cleft. The mean spark life time was obtained to 9.8 ms.

are additionally several channels used which are treated in a stochastic way, but the diadic cleft was not modelled in a spatially resolved way. Greenstein et al. [6] considered some spatial structure and concentration gradients by introducing compartments for each of 4 pairs of one LCC and 5 RyRs, simulated each channel as Markov chain and concentrations dynamically with rate equations. Our approach is more flexible with respect to channel numbers, more detailed with respect to spatial structure and computationally more efficient, since integrate only the differential equation for c_{JSR} along with the Markov simulations.

There is also a wide class of papers dealing only to modell the diadic cleft. For example Tanskanen et al. [18], treat the diadic cleft not only spatially resolved, but also takes explicitly into account the protein structure in the diadic cleft and membrane buffers. The channel gating and the flux through open channels is treated stochastically. A similar approch was presented in Koh et al. [11]. Nevertheless this models are very time expensive and not suitable for whole cell simulations.

The introduced approach opens up new possibilities to study the influence of altering channel properties in the framework of the whole cell. Interesting phenomena in this background are for instance channel phosphorylation or cooperative RyR gating. Channel phoshorylation can be mediated by the cAMP/PKA pathway due to β-adrenergic stimulation. This leads to an increase of the flux through RyRs and LCCs [7, 10, 13]. Another example would be the effect of cooperating RyRs. The RyR are linked with their neighbours by a protein (FKBP), which couples the gating of the channels [12, 19].

5. Appendix

5.1. *Ca²⁺ Profile for More Than One Open Channel and with Varying Bulk Concentration*

For solving the problem with N channels, it is necessary to introduce channel related concentrations c_i. $c_i(r)$ describes the part of the Ca²⁺ concentration at position r, that is produced by the ith channel. Furthermore, we introduce $\alpha_{i,j} = \frac{c_i(r_j)}{c_i(r_i)}$. For the Ca²⁺ bulk concentration, we have to determine the parameter β by $c_{bulk} - c_{rest} = \beta \sum_i J_i$. For an understandable derivation we just show only the case of open RyRs, an extension for LCCs is straight forward. In order to solve the problem, we obtained following system of linear equations:

$$\overline{c_1} = \kappa_1 g \left(c_{JSR} - c_{bulk} - \overline{c_1} - \sum_{j \neq 1}^{N} \alpha_{j,1} \overline{c_j} \right)$$

$$\overline{c_2} = \kappa_2 g \left(c_{JSR} - c_{bulk} - \overline{c_2} - \sum_{j \neq 2}^{N} \alpha_{j,2} \overline{c_j} \right)$$

$$\dots$$

$$\overline{c_N} = \kappa_N g \left(c_{JSR} - c_{bulk} - \overline{c_N} - \sum_{j \neq N}^{N} \alpha_{j,N} \overline{c_j} \right)$$

$$c_{bulk} - c_{rest} = \beta \sum_{i}^{open} J_i \tag{17}$$

In the maximal case that all channels are open we end up with a 21x21 matrix, which can be solved numerically fast.

5.2. *Green's Function*

We assumed the myocyte as a cylinder with $r \in [0, R]$ and $z \in [-\frac{L}{2}, \frac{L}{2}]$. To solve the system of PDE (15), the Green's function $\tilde{G}$ must in general take the form:

$$\tilde{G} = \sum_{k,m=0}^{\infty} \sum_{n=0}^{\infty} \alpha_{k,m,n} B_0(x(k)\frac{r}{R}) B_0(x(k)\frac{r'}{R}) cos(m\pi\frac{z}{L}) cos(m\pi\frac{z'}{L}) cos(n(\varphi - \varphi'))$$

$$\tag{18}$$

with the bessel function of the first kind, $0th$ order $B_0(r)$ which satiesfies the Helmholtz equation. The eigenvalues $x(k)$ are determined by the boundary condition at R (the cell radius) of no fluxes:

$$\frac{\partial B_0(x(k)\frac{r}{R})}{\partial r}\Big|_{r=R} = 0 \tag{19}$$

In the particular case of one δ-source located at $r' = 0$ and $z' = 0$ Eq.(18) simplifies to

$$\tilde{G} = \sum_{k,m=0}^{\infty} \tilde{\alpha}_{k,m} B_0(x(k)\frac{r}{R}) cos(2m\pi\frac{z}{L}) \tag{20}$$

After some steps with using Eq.(16) and the fact that we have only a single δ-channel at the centre of the cell we arrive at (for more details, see [16]):

$$\begin{pmatrix} \delta c \\ \delta M \\ \delta T \end{pmatrix}(t) = \sum_{k=0}^{\infty} \sum_{m=0}^{\infty} B_0(x(k)\frac{r}{R}) cos(2m\pi\frac{z}{L}) \begin{pmatrix} \chi_1^{(k,m)} \\ \chi_2^{(k,m)} \\ \chi_3^{(k,m)} \end{pmatrix} (\vec{r'} = \vec{0}, t) \tag{21}$$

The response functions $\vec{\chi}^{(k,m)}$ include the time integration over the time dependent channel flux J and includes the interaction with the buffers as well as the pumping to the SR:

$$\begin{pmatrix} \chi_1^{(k,m)} \\ \chi_2^{(k,m)} \\ \chi_3^{(k,m)} \end{pmatrix} = \sum_{i=0}^{3} \frac{1}{N_{k,m}} \int_0^t d\tau\, J(\tau) e^{-s_i(t-\tau)} \frac{adj(M_{km}(s_i))}{\partial |M_{km}|/\partial s|_{s=s_i}} \begin{pmatrix} 1 \\ 0 \\ 0 \end{pmatrix} \tag{22}$$

with the normalization factor N (for $k, m \neq 0$):

$$N_{k,m} = \int_0^R \int_0^{2\pi} \int_{-\frac{L}{2}}^{\frac{L}{2}} dr\, d\varphi\, dz\, r\, B_0(x(k)\frac{r}{R}) cos(2m\pi\frac{z}{L}) = \frac{\pi R^2 L}{2} \tag{23}$$

The s_i of the response function can be determined by $det(M_{km}) = 0$. M is given by:
$$M_{km}=$$

$$\begin{pmatrix} -D_c x(k)^2 - s - \rho_{SERCA} - k_+^M M_0 - k_+^T T_0 & -k_+^M c_0 & -k_+^T c_0 \\ -k_+^M M_0 & -D_M x(k)^2 - s - k_+^M c_0 & 0 \\ -k_+^T T_0 & 0 & -s - k_+^T c_0 \end{pmatrix}$$

5.3. *Parameters*

Table 1 RyR parameters   LCC parameters

Parameter	Value	Parameter	Value
g	$10\mu m^3 s^{-1}$	J_L	$0.913\mu m^3 s^{-1}$
k_{RO}	$0.015\mu M^{-2}s^{-1}$	V_L	$-2mV$
k_{OR}	$60s^{-1}$	$\Delta V_L r$	$7mV$
k_{IO}	$5s^{-1}$	ϕ_L	6.667
k_{OI}	$0.5\mu M^{-1}s^{-1}$	t_L	$0.0033s$
k_{IR}	$5s^{-1}$	τ_L	$0.65s$
k_{RI}	$0.5\mu M^{-1}s^{-1}$	K_L	$0.22\mu M$
k_{IR2}	$60s^{-1}$	a	0.0625
k_{RI2}	$0.015\mu M^{-2}s^{-1}$	b	14

Table 2 Other parameters

Parameter	Value	Parameter	Value
c_{ext}	$1000\mu M$	k_M^+	$100\mu M^{-1}s^{-1}$
c_{NSR}	$700\mu M$	k_M^-	$38s^{-1}$
D_c	$100\mu m^2 s^{-1}$	k_T^+	$40\mu M^{-1}s^{-1}$
B_{CSQN}	$400\mu M$	k_T^-	$40s^{-1}$
K_c	$600\mu M$	B_M	$24\mu M$
τ_{refill}	$7.5ms$	B_T	$70\mu M$
ρ_{SERCA}	$390s^{-1}$		

References

[1] Brochet, D.X.P., Yang, D.M., Di Maio, A., Lederer, J., Franzini-Armstrong, C., Cheng, H.P. Ca^{2+} blinks: rapid nanoscopic store calcium signaling. *Proc. Natl. Acad. Sci.USA.*, 102:3099–3104, 2005.

[2] Cheng, H., Fill, M., Valdivia, H., Lederer, W.J., Models of Ca^{2+} release channel adaption. *Science*, 267:2009–2011, 1995.

[3] Chen-Izu, Y., McCulle, S.L., Ward, C.W., Soeller, C., Allen, B.M., Rabang, C., Cannell, M.B., Balke, C.W., Izu, L. Three-Dimensional Distribution of Ryanodine Receptor Clusters in Cardiac Myocytes. *Biophy. J.*, 91:1–13, 2006.

[4] Cleeman, L., Wang, W., Morad, M. Two-dimensional confocal images of organization, density, and gating of focal Ca^{2+} release sites in rat cardiac myocytes. *Proc. Natl. Acad. Sci.USA.*, 95:10984–10989, 1998.

[5] Gillespie, D.T., Exact stochastic simulation of coupled chemical reactions. *J. Phys. Chem.*, 81: 2340–2361, 1977.

[6] Greenstein, J.L., Winslow, R.L., An integrative model of the cardiac ventricular myocyte incorporating local control of Ca^{2+} release. *Biophy. J.*, 83:2918–2945, 2002.

[7] Hain, J., Onoue, H., Mayrleitner, M., Fleischer, S., Schindler, H., Phosphorylation modulates the function of the calcium release channel of sarcoplasmic reticulum from cardiac muscle. *J.Biol.Chem.*, 270:2074–2081, 1995.

[8] Hake, J., Lines, G.T., Stochastic binding of Ca^{2+} ions in the dyadic cleft; continious vs. Random Walk description of diffusion. *Biophys. J.*, 94:4184–4201, 2008.

[9] Hinch, R., Greenstein, J.L., Tanskanen, A.J., Xu, L., Winslow, R.L., A simplified local control model of calcium-induced calcium release in cardiac ventricular myocytes. *Biophy. J.*, 85:3723–3736, 2004.

[10] Kamp, T.J., Hell, J.W., Regulation of cardiac L-type calcium channels by protein kinase A and protein kinase C. *Circ. Res.*, 87:1095–1102, 2000.

[11] Koh, X., Srinivasan, B., Ching, H.S., Levchenko, A., A 3D Monte Carlo analysis of the role of dyadic space geometry in spark generation. *Biophys. J.*, 90:1999–2014, 2006.

[12] Marx, S.O., Gaburjakova, J., Gaburjakova, M., Hendrikson, C., Ondrias, K., Marks, A.R., Coupled gating between cardiac Ca^{2+} release channels (ryanodine receptors). *Circ.Res.*, 88:1151–1158, 2001.

[13] Reiken, S., Garbujakova, M., Guatimosin, S., Gomez, A.M., D'Armiento, J., Burkhoff, D., Wang, J., Vassort, G., Lederer, W.J., Marks, A.R., Protein kinase A phosphorylation of the cardiac calcium release channel (ryanodine receptor) in normal and failing hearts. Role of phosphotases and response to isoprotenerol. *J.Biol.Chem.*, 278:444–453, 2003.

[14] Restrepo, J.G., Weiss, J.N., Karma, A., Calsequestrin-Mediated Mechanism for Cellular Calcium Transient Alternans. *Biophy. J.*, 95:3767–3789, 2008.

[15] Rüdiger, S., Shuai, J.W., Huisinga, W., Nagaiah, C., Warnecke, G., Parker, I., Falcke, M., Hybrid stochastic and deterministic simulations of calcium blips. *Biophy. J.*, 93:1847–1857, 2007.

[16] Skupin, A., Falcke, M., The role of IP_3R clustering in Ca^{2+} signaling. *Genome Informatics*, 20:15–24, 2008.

[17] Soeller, C., Cannel, M.B., Numerical simulation of local calcium movements during L-type calcium channel gating in the cardiac diad. *Biophy. J.*, 73:97–111, 1997.

[18] Tanskanen, A.J., Greenstein, J.L., Chen, A., Sun, S.X., Winslow, R.L., Protein geometry and placement in the cardiac dyad influence macroscopic properties of calcium-induced calcium release. *Biophy. J.*, 92:3379–3396, 2007.

[19] Wagenknecht, T., Badermacher, M., Grassucci, R., Berkowitz, J., Xin, H.B., Fleischer,

S., Locations of calmodulin and FK506-binding protein on the three-dimensional architecture of the sceletal muscle ryandine receptor. *J.Biol.Chem.*, 272:32463–32471, 1997.

[20] Wang, S.Q., Song, L.S., Lakkata, E.G., Cheng, H., Ca^{2+} signalling between single L-type Ca^{2+} channels and ryanodine receptors in heart cells. *Nature*, 410:592–596, 2001.

[21] Wang, S.Q., Stern, M.D., Rios, E., Cheng, H., The quantal nature of Ca^{2+} sparks and in situ operation of the ryanodine receptor array in cardiac cells. *Proc. Natl. Acad. Sci.USA.*, 101:3979–3984, 2003.

[22] Wier, W.G., Egan, T.M., Lopez-Lopez, J.R., Balke, C.W., Local control of excitation-contraction coupling in rat heart cells. *J.Physiol.*, 474:463–471, 1994.

[23] Williams, G.S.B., Huertas, A.H., Sobie, E.A., Jafri, M.S., Smith, G.D., A Probabilty density approach to modeling local control of calcium-induced calcium release in cardiac myocytes. *Biophy. J.*, 92:2311–2328, 2007.

CO-EVOLUTION OF METABOLISM AND PROTEIN SEQUENCES

MORITZ SCHÜTTE[1] NIELS KLITGORD[2]
schuette@mpimp-golm.mpg.de niels@bu.edu

DANIEL SEGRÈ[2,3] OLIVER EBENHÖH[1,4,5,6]
dsegre@bu.edu ebenhoeh@abdn.ac.uk

[1] *Max Planck Institute of Molecular Plant Physiology, Am Mühlenberg 1, 14476 Potsdam–Golm, Germany*
[2] *Boston University, Bioinformatics Program, 24 Cummington Street, Boston, MA 02215, USA*
[3] *Boston University, Departments of Biology and Biomedical Engineering, 24 Cummington Street, Boston, MA 02215, USA*
[4] *Potsdam University, Institute of Biochemistry and Biology, Karl–Liebknecht–Straße 24–25, 14476 Potsdam–Golm, Germany*
[5] *Institute for Complex Systems and Mathematical Biology, University of Aberdeen, Aberdeen, AB24 3UE, UK*
[6] *Institute of Medical Sciences, Foresterhill, University of Aberdeen, Aberdeen, AB25 2ZD, UK*

The set of chemicals producible and usable by metabolic pathways must have evolved in parallel with the enzymes that catalyze them. One implication of this common historical path should be a correspondence between the innovation steps that gradually added new metabolic reactions to the biosphere-level biochemical toolkit, and the gradual sequence changes that must have slowly shaped the corresponding enzyme structures. However, global signatures of a long-term co-evolution have not been identified. Here we search for such signatures by computing correlations between inter-reaction distances on a metabolic network, and sequence distances of the corresponding enzyme proteins. We perform our calculations using the set of all known metabolic reactions, available from the KEGG database. Reaction-reaction distance on the metabolic network is computed as the length of the shortest path on a projection of the metabolic network, in which nodes are reactions and edges indicate whether two reactions share a common metabolite, after removal of cofactors. Estimating the distance between enzyme sequences in a meaningful way requires some special care: for each enzyme commission (EC) number, we select from KEGG a consensus set of protein sequences using the cluster of orthologous groups of proteins (COG) database. We define the evolutionary distance between protein sequences as an asymmetric transition probability between two enzymes, derived from the corresponding pair-wise BLAST scores. By comparing the distances between sequences to the minimal distances on the metabolic reaction graph, we find a small but statistically significant correlation between the two measures. This suggests that the evolutionary walk in enzyme sequence space has locally mirrored, to some extent, the gradual expansion of metabolism.

Keywords: metabolism; networks; evolution; protein sequence; enzyme; EC number; KEGG; COG.

1. Introduction

The evolutionary walk from an early proto-metabolism to the current biochemical pathways must have been shaped by innovations concurrently involving enzymes and chemical compounds [13]. While it is generally assumed that todays enzymes have evolved from a few ancestors that were able to catalyze the first reactions, a clear correspondence between the evolution of metabolic functions and their catalyzing enzymes remains to be established. The evolution of metabolic pathways has been addressed by several competing models, including the patchwork model [6, 27], and models of forward [3] and retrograde [4] evolution. In addition, several studies have addressed the relation between sequence homology and protein function. These studies have been widely used for the prediction of protein functions [19, 24, 26] associated with newly sequenced genes and for the analysis of relations between sequences and functions in Gene Ontology terms [7].

To date, there exists only few studies of evolutionary relation between enzyme sequence homology and distance on the metabolic reaction network. Most of these studies are restricted to networks of single organisms. These works show a possible link between homology and metabolic network distance. For example in *E.coli* [15, 18], it was found that homologous protein sequences are more likely to be found in close vicinity on the metabolic network than what expected by chance and the same trend has been confirmed for protein-protein interactions [5]. Similarly, in studies of yeast [25] a link has been found between the metabolic network structure and enzymatic evolution. Since single enzymes or even entire operons have been copied and changed to fulfill new functions, high promiscuity in the locality of catalyzing enzymes complicates the search for evolutionary relations [11, 20, 21].

Here, we test the hypothesis that the global scope of metabolism has evolved in parallel with the enzymes that make up the network. To investigate this hypothesis, we have taken a large-scale approach using the entire set of reactions from the KEGG [8–10] reaction database. If our hypothesis holds true, then we expect to find that a similarity between protein sequences should be reflected by a closeness in the metabolic reaction network. In order to test our hypothesis we define different measures of sequence distances, both symmetric and asymmetric, based on a reciprocal pair-wise BLAST analysis. Asymmetry becomes important if two sequences of different lengths are compared, because it is more likely that a shorter sequence is the evolutionary child of a longer sequence, than the other way around. These distances are then compared with the distances on an enzyme-enzyme network of chemical reactions that we construct from the KEGG database. As KEGG provides multiple protein sequences per reaction, we first choose a consensus sequence set based on the cluster of orthologous groups of proteins (COG) database [22, 23] that greatly reduces the sequence space we must analyze. Our analysis supports our hypothesis, showing that enzymes that are close in the reaction network are enriched for sequence similarity. The observed trend is small but significant against a simulated control.

2. Protein Sequence Distances and Enzyme Distances

To address our hypothesis on a large scale, we use the entirety of reactions from the KEGG database. We construct a reaction-centric network where reactions are nodes and two reactions are linked if they involve a common metabolite. Some metabolites, such as cofactors, participate in a variety of reactions and thus produce short-cuts that do not carry actual fluxes. To account for this, we extract the cofactor pairs ATP/ADP, NAD^+/NADH, $NADP^+$/NADPH, and CoA/Acetyl-CoA from the reactions where they appear as pairs [5]. Furthermore, we delete highly abundant molecules, H_2O, H^+, and O_2, which appear in more than 500 reactions, from the reaction set. Every reaction is catalyzed by one or more enzymes given in terms of enzyme commission (EC) numbers, exceptions in the reaction set are spontaneous reactions or those for which the catalyst is not known. We link the reactions to the enzyme sequence using the EC numbers. Such, we transform the reaction network to an enzyme-enzyme network. As we only know sequences for roughly half of all EC numbers we still use all reaction links but only calculate shortest paths between enzymes for which we know the sequences of starting and ending enzymes. For this purpose we use the Dijkstra algorithm [1]. Intermediate reactions without catalyzing enzyme, like spontaneous reactions, or with an enzyme without known sequence, are still counted as a step in the distance.

The distance in protein sequence space requires more elaboration. We use the Blastp in the *bl2seq* program to obtain pair-wise comparisons between sequences [28]. A rather intuitive measure obtained from blast is the *identity* measure counting how many residues on a certain aligned fragment are identical between the two sequences. To get a comparable scoring measure we need to normalize the identities by the sequence lengths. It is important to note that we are interested in interpreting such a measure as an estimate of the probability that one sequence has evolved from the other. Since asymmetry can play a crucial role in the transition from one sequence to the other [16], we will take into account not only the total length, but also the difference between the two lengths. We define $\mathcal{I} = \sum_i I_i$ as the sum of the identities of every found alignment. Based on this measure, we define three different estimates of the probability Γ_{AB} that a sequence B (with length L_B) has evolved from a sequence A (with length L_A):

$$\Gamma_{AB}^{(\mathrm{id})} = 1 - \frac{2 \cdot \mathcal{I}}{L_A + L_B}, \tag{1}$$

$$\Gamma_{AB}^{(\mathrm{as1})} = 1 - \frac{2 \cdot \mathcal{I}}{L_A + L_B} \left[1 + \frac{(L_A - L_B)}{(L_A + L_B)} \right], \tag{2}$$

$$\Gamma_{AB}^{(\mathrm{as2})} = 1 - \frac{\mathcal{I}}{L_B}. \tag{3}$$

The second measure follows from a Taylor expansion of Eq. (1). We consider the length difference as a small correction to the mean of the lengths: $(L_A + L_B)/2 \implies \bar{L} - \Delta L/2$: $(\bar{L} - \Delta L/2)^{-1} \approx 1/\bar{L} + \Delta L/2\bar{L}^2$. If the directionality of a linear pathway is known, an asymmetric distance like Eqs. (2)–(3) could in principle be used to test for retrograde or forward evolution [3, 4]. In addition to Eqs. (1)–(3) we use the *score* values of the best hit obtained from BLAST and normalize it by the score of the sequence with itself:

$$\Gamma^{(sc)}_{AB} = 1 - \frac{2 \cdot score(A, B)}{score(A, A) + score(B, B)} \,.$$

(4)

To evaluate the utility of these proposed measures, we tested how they perform on simple pairs of sequences, computed so as to simulate a gradual transition from exact identity to large differences, see Fig. 1. For the purposes of our simulation, we chose one random but rather long (839 amino acids) test sequence (*glo:Glov_1829 integral membrane sensor signal transduction histidine kinase (EC:2.7.13.3)*). These experiments simulate some of the possible scenarios of sequence changes during enzyme evolution. In the first two experiments, we took a copy of the sequence and iteratively cut off amino acids from either start or end of the copy. This reduced sequence is then blasted against the original one, Figs. 1(a) and (b). In experiment three we shuffled increasingly more amino acids in the copy and blasted against the

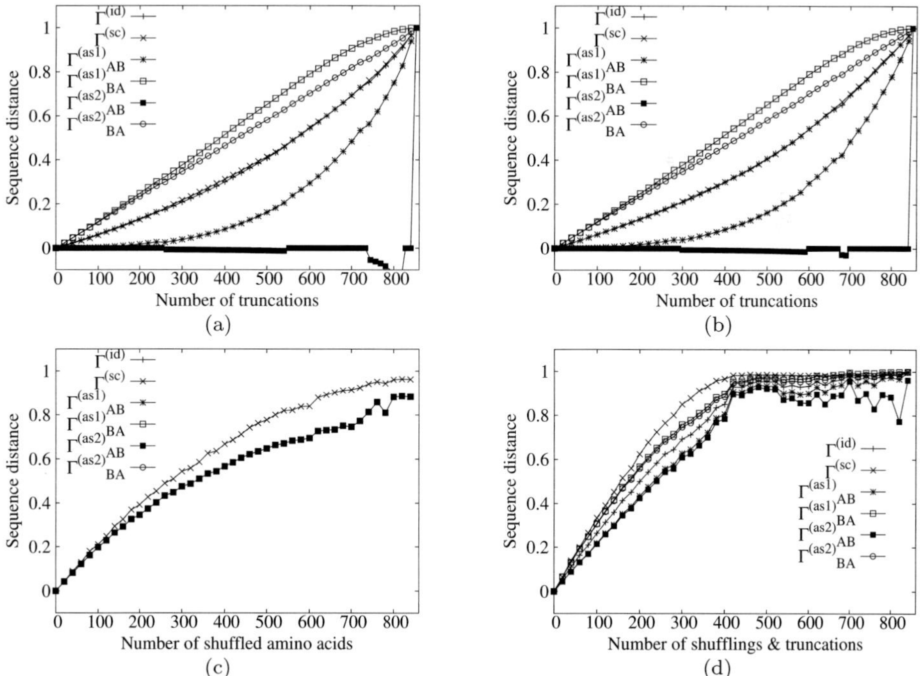

Fig. 1: Simulation of four evolutionary scenarios. We take a test sequence, change it and blast it against an original copy. In (a) and (b) we iteratively delete amino acids from the start (a) or end (b) of the test sequence. (c): The amino acids are increasingly shuffled. (d) combines (a) and (c): we shuffle and truncate the sequence.

original sequence, (c). The last experiment, (d), is a combination of shuffling and length reduction. Here, we observe an edge at around 0.9 distance. Below this value the behavior seems random.

These experiments show that $\Gamma_{AB}^{(as2)}$, Eq. (3), is not an appropriate measure. Specifically, in the first two experiments this measure reaches negative values, resulting from the fact that we sum all identities. Because the aligned fragments become very small, it is likely that the same fragment is found twice or that an overlap between two matches is found in the original copy. Since the measure is only normalized by its own length, this may result in negative values. A similar artifact is also observed in experiment four, where the measure shows a false positive agreement when sequences have been highly shuffled, and greatly truncated. Thus this measure will not be used for further investigations.

3. Consensus Sequence Set

We characterize an enzyme by two features: its sequence, and its function. These features can, to some degree, vary independently: One enzyme may have multiple functions, or conversely, one specific function can be performed by multiple sequences [2]. Our goal is to investigate whether the evolutionary distance of sequences relates to their functional distance as determined by the metabolic reaction network where the reactions are defined in terms of EC numbers. Since each EC number can be associated with a rather large number of sequences, we define a consensus set of sequences serving as representatives of the specific function. In total, the KEGG database contains approximately 750000 sequences that contain one or more EC numbers in their description. As can be seen in Fig. 2(a), the distribution

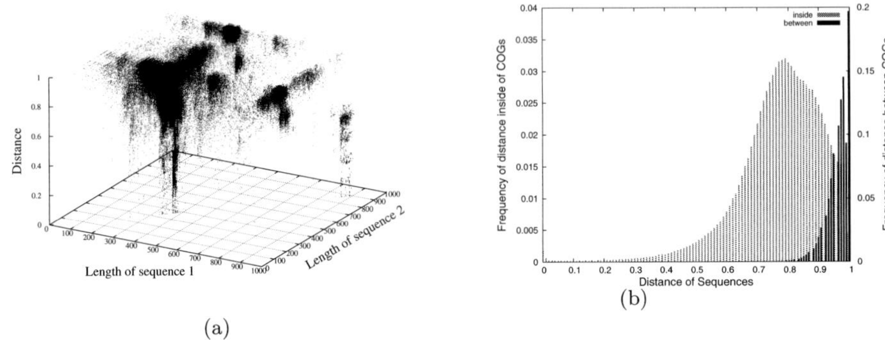

(a)   (b)

Fig. 2: (a) All pair-wise blasts of the alcohol dehydrogenase (EC 1.1.1.1) sequences scored by the distance Eq. (1). Two apparent clusters of small distance are observable around length 300 and 900. (b) Benchmark of COG inner distance versus the distances between representatives of each COG. The cut-off of 0.9 to differ seems reasonable.

of the number of sequences per EC number can vary quite a bit in KEGG. This has been greatly reduced in our consensus sequence set. For example, the sequence by sequence distances for alcohol dehydrogenase, EC 1.1.1.1, are shown in Fig. 2(a). The lengths vary by a factor three and even the sequences of similar lengths need not at all be similar. Even the set of 188 alcohol dehydrogenase sequences that are all 350 amino acids in length varies from completely identical to completely different with a mean of $\Gamma^{(\mathrm{id})} = 0.72 \pm 0.12$. One can observe several distinct clusters of high similarity in Fig. 2(a).

To simplify our task of selecting representative sequences, we utilize the COG database [22, 23] that clusters proteins by their function based on homology. Every COG contains between a few and a few hundred sequences that are related by a duplication or speciation event. We pick the longest sequence of every COG as the representative of this COG [12]. In order to cover as many as possible EC numbers we cluster the remaining KEGG sequences by a very simple procedure. We group the sequences by EC number and perform all pairwise blasts dropping those sequences that have a 0.9 or higher distance according to Eq. (1). This loose cut-off is justifiable using the COGs as a benchmark. We calculated all inner-COG distances and compared them with the distances between the representatives of every COG, see Fig. 2(b). A second argument for this cutoff comes through the

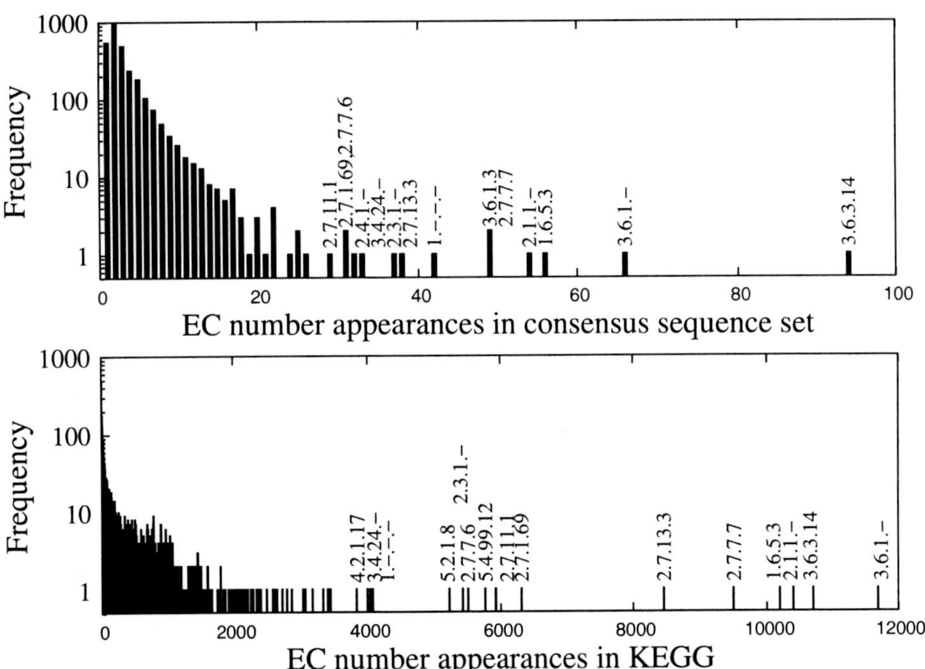

Fig. 3: Distribution of EC numbers how often they appear in descriptions of different protein sequences. Top in the consensus set, bottom in the KEGG sequences.

result of Fig. 1 (d) where all curves show a somewhat random behavior for distances larger than 0.9.

Following this procedure we obtain a consensus set of 8123 sequences coding for 2821 EC numbers. Fig. 3 shows a histogram of the frequency of a certain EC number with that it appears in descriptions of different sequences. The top bar graph shows the distribution in the final consensus set and the bottom one in the starting set from all KEGG sequences. There is a good agreement between the ranking by EC number with Pearson correlation 0.81 and Spearman Rank correlation 0.53.

4. Correlation of Network and Sequence Distances

We use the previously defined consensus sequence set to analyze a relationship between distance on the enzyme-enzyme graph and the sequences of the enzymes. We use a sample of 4.8 million shortest paths which all start and end with a sequenced enzyme.

Figure 4(a) shows a boxplot of the correlation using the measure $\Gamma^{(\mathrm{id})}$, Eq. (1). We see a highly significant but small correlation that is mirrored by a trend seen in the outliers where similar enzymes tend to be closer in the network. In order to test the results against the null hypothesis that they appeared by chance, we calculated the p-value which is based on the sample size. We performed a second control calculation utilizing a permutation test where we shuffled the sequence distances to generate a random set. For this control simulation the correlation is completely lost and we observe high p-values, see Tab. 1.

To further quantify the observation, we analyze the results sorting them by particular path-lengths, Fig. 4(b). Neighboring enzymes tend to show higher similarity on the sequence level. For enzyme-enzyme distance 1, the relative proportion of distances below 0.8 and 0.7 is enriched. The bar on distance 8 is the highest but it represents only a sample of four similar enzymes of 47.

Table 1: Comparison of correlations between enzyme sequence distances and distances obtained form the enzyme-enzyme graph. In the control measurement we shuffle the enzyme distance matrix and repeat the simulation. The distances in the sequence space were calculated with the measures described in section 2 (sample size: 4.8 million shortest paths). Although the correlations are very low, they are highly significant in comparison with the control data.

measure	$\Gamma^{(\mathrm{id})}_{AB}$	control to $\Gamma^{(\mathrm{id})}_{AB}$	$\Gamma^{(\mathrm{as1})}_{AB}$	control to $\Gamma^{(\mathrm{as1})}_{AB}$	$\Gamma^{(\mathrm{sc})}_{AB}$	control to $\Gamma^{(\mathrm{sc})}_{AB}$
correlation	0.0127	0.0002	0.0170	-0.0003	0.0035	0.0004
p-value	10^{-171}	0.7270	10^{-304}	0.4931	10^{-14}	0.4189

Table 1 compares the results for different measures of sequence similarity. The asymmetric measure Eq. (2) yields the highest correlation and significance in calculating the sequence distance. For every enzyme pair we chose the more similar value of the asymmetric sequence distance $\Gamma^{(\mathrm{as1})}$.

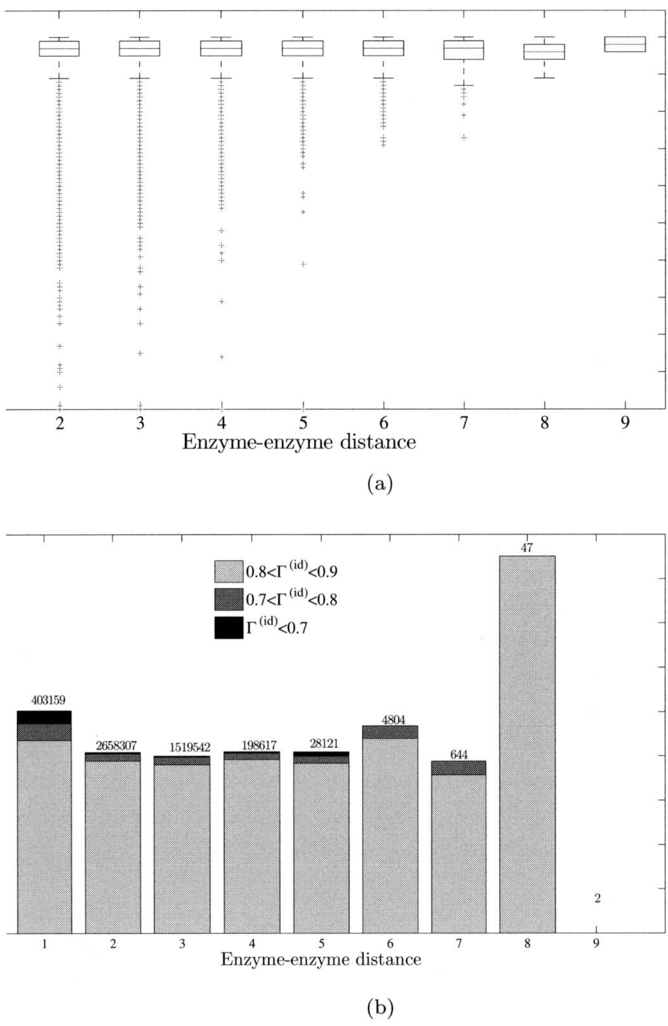

Fig. 4: (a) Boxplot for the correlation between distances on the enzyme-enzyme graph and the measure $\Gamma^{(\mathrm{id})}$ for the sequences. The correlation is very low, 0.0127, but significant, see Tab. 1. (b) Sequence distances sorted by particular enzyme-enzyme distances. The plot shows the fraction of enzyme sequence pairs within a certain distance of $\Gamma^{(\mathrm{id})}$ compared to all enzymes found in the distance on the network. The small numbers on top of the bars represent the total number of enzymes found in the particular distance. For neighboring enzymes we observe a higher fraction of enzymes with similar sequences.

5. Conclusion

We have investigated the evolutionary relation of enzyme sequences and their distance in the metabolic network on a large-scale using a consensus sequence set from the entire KEGG database. However, the choice of the consensus set is strongly biased by two aspects of the used database: the choice of sequenced organisms and the accuracy in investigating proteins. A large number of redundant sequences is due to the variety of organisms whose proteomes are sequenced, and whose function was assigned via homology. The sequences in the consensus set come from 27 animals, 6 plants, 21 fungi, 535 bacteria, 51 archea and 17 protists and these result in 1395, 274, 585, 4974, 598, 297 sequences from the particular kingdoms. The variability in plant-specific enzymes might be underestimated as only a few model plants are well investigated. For the carbon-fixating enzyme RuBisCO (EC 4.1.1.39) we obtain only two different sequences from bacteria *Synechocystis* and *Anabaena*. The majority of organisms are bacteria for which lateral gene transfer is an important factor [14, 17]. By the use of the COG database we might neglect this possibility of sequence change. The second bias appears through the way proteins are investigated. As an example we examine ATP synthase, EC 3.6.3.14. This protein is the second most abundant in KEGG and the most abundant in the consensus set, Fig. 3. It catalyzes only one reaction, $ATP + H_2O + H_{in}^+ \rightleftharpoons ADP + phosphate + H_{out}^+$. This reaction is essential in most organisms and frequently investigated. We thus capture the variability of sequences for known enzymes but do not grasp it for less known ones.

We have observed a weak correlation between enzymes that are neighbors in the graph representing metabolism, and the corresponding sequences. Our finding extends to an all-organism level results previously obtained for single organisms [15] and using protein-protein interactions [5]. The correlation detected indicates a certain degree of co-evolution between the topology of metabolism and its enzyme capabilities. We envisage that future simulations of the evolutionary expansion of metabolism, possibly employing our proposed asymmetric measure of sequence distance, could shed more insight into the nature of this correlation.

6. Acknowledgments

We acknowledge financial support from the International Research Training Group *Genomics and Systems Biology of Molecular Networks* IRTG 1360 (MS), the German Federal Ministry of Education and Research, Systems Biology Research Initiative *Go*FORSYS, the Scottish University Life Science Alliance SULSA (OE), the NASA Astrobiology Institute, and the US Department of Energy (NK, DS).

References

[1] Dijkstra, E.W., A note on two problems in connexion with graphs, *Numerische Mathematik*, 1: 269–271, 1959.

[2] Galperin, M.Y., Walker, D.R., Koonin, E.V., Analogous enzymes: independent inventions in enzyme evolution, *Genome Res.*, 8: 779–790, 1998.

[3] Granick, S., Speculations on the origins and evolution of photosynthesis, *Ann N Y Acad. Sci.*, 69: 292–308, 1957.

[4] Horowitz, N.H., On the evolution of biochemical syntheses, *PNAS*, 31(6): 153–157, 1945.

[5] Huthmacher, C., Gille, C., Holzhütter, H.G., A computational analysis of protein interactions in metabolic networks reveals novel enzyme pairs potentially involved in metabolic channeling, *J. Theor. Biol.*, 252: 456–464, 2008.

[6] Jensen, R.A., Enzyme recruitment in evolution of new function, *Annu. Rev. Microbiol.*, 30: 409–435, 1976.

[7] Joshi, T., Xu, D., Quantitative assessment of relationship between sequence similarity and function similarity, *BMC Genomics*, 8: 222, 2007.

[8] Kanehisa, M., Goto, S., KEGG: Kyoto Encyclopedia of Genes and Genomes, *Nucleic Acids Res.*, 28: 27–30, 2000.

[9] Kanehisa, M., Goto, S., Hattori, M., Aoki-Kinoshita, K.F., Itoh, M., Kawashima, S., Katayama, T., Araki, M., Hirakawa, M., From genomics to chemical genomics: new developments in KEGG, *Nucleic Acids Res.*, 34: 354–357, 2006.

[10] Kanehisa, M., Araki, M., Goto, S., Hattori, M., Hirakawa, M., Itoh, M., Katayama, T., Kawashima, S., Okuda, S., Tokimatsu, T., Yamanishi, Y., KEGG for linking genomes to life and the environment, *Nucleic Acids Res.*, 36: 480–484, 2008.

[11] Khersonsky, O., Roodveldt, C., Tawfik, D.S., Enzyme promiscuity: evolutionary and mechanistic aspects, *Curr Opin Chem Biol.*, 5: 498–508, 2006.

[12] Krause, A., Stoye, J., Vingron, M., Large scale hierarchical clustering of protein sequences, *BMC Bioinformatics*, 6: 15, 2005.

[13] Lazcano, A., Miller, S.L., On the origin of metabolic pathways, *J. Mol. Evol.*, 49: 424–431, 1999.

[14] Lercher, M.J., Pàl, C., Integration of horizontally transferred genes into regulatory interaction networks takes many million years, *Mol. Biol. Evol.*, 25(3): 559–567, 2008.

[15] Light, S., Kraulis, P., Network analysis of metabolic enzyme evolution in *Escherichia coli*, *BMC Bioinformatics*, 5: 15, 2004.

[16] Notebaart, R.A., Kensche, P.R., Huynen, M.A., Dutilh, B.E., Asymmetric relationships between proteins shape genome evolution, *Genome Biol.*, 10: R19, 2009.

[17] Pàl, C., Papp, B., Lercher, M.J., Adaptive evolution of bacterial metabolic networks by horizontal gene transfer, *Nat Genet.*, 17(12): 1372–1375, 2005.

[18] Rison, S.C., Teichmann, S.A., Thornton, J.M., Homology, pathway distance and chromosomal localization of the small molecule metabolism enzymes in *Escherichia coli*, *J. Mol. Biol.*, 318(3): 911–932, 2002.

[19] Rost, B., Enzyme function less conserved than anticipated, *J. Mol. Biol.*, 26: 595–608, 2002.

[20] Schmidt, S., Sunyaev, S., Bork, P., Dandekar, T., Metabolites: a helping hand for pathway evolution?, *Trends Biochem Sci.*, 6: 336–341, 2003.

[21] Spirin, V., Gelfand, M.S., Mironov, A.A., Mirny, L.A., A metabolic network in the evolutionary context: multiscale structure and modularity, *Proc Natl Acad Sci*, 103(23): 8774–8779, 2006.

[22] Tatusov, R.L., Koonin, E.V., Lipman, D.J., A genomic perspective on protein families, *Science*, 278: 631–637, 1997.

[23] Tatusov, R.L., Fedorova, N.D., Jackson, J.D., Jacobs, A.R., Kiryutin, B., Koonin, E.V., Krylov, D.M., Mazumder, R., Mekhedov, S.L., Nikolskaya, A.N., Rao, B.S., Smirnov, S., Sverdlov, A.V., Vasudevan, S., Wolf, Y.I., Yin, J.J., Natale, D.A., The COG database: an updated version includes eukaryotes, *BMC Bioinformatics*, 4: 41, 2003.

[24] Tian, W., Skolnick, J., How well is enzyme function conserved as a function of pairwise sequence identity?, *J. Mol. Biol.*, 31: 863–882, 2003.

[25] Vitkup, D., Kharchenko, P., Wagner, A., Influence of metabolic network structure and function on enzyme evolution, *Genome Biology*, 7: R39, 2006.

[26] Weinhold, N., Sander, O., Domingues, F.S., Lengauer, T., Sommer, I., Local function conservation in sequence and structure space, *PLoS Comput Biol*, 4(7): e1000105, 2008.

[27] Ycas, M., On earlier states of the biochemical system, *J. Theor. Biol.*, 44: 145–160, 1974.

[28] `http://blast.ncbi.nlm.nih.gov/`

CHARACTERIZATION AND CLASSIFICATION OF ADVERSE DRUG INTERACTIONS

MASATAKA TAKARABE[1]   DAICHI SHIGEMIZU[1]   MASAAKI KOTERA[1]
takarabe@kuicr.kyoto-u.ac.jp daichi@kuicr.kyoto-u.ac.jp   kot@kuicr.kyoto-u.ac.jp

SUSUMU GOTO[1]   MINORU KANEHISA[1,2]
goto@kuicr.kyoto-u.ac.jp   kanehisa@kuicr.kyoto-u.ac.jp

[1] *Bioinformatics Center, Institute for Chemical Research, Kyoto University, Uji, Kyoto 611-0011, Japan*
[2] *Human Genome Center, Institute of Medical Science, University of Tokyo, Minato-ku, Tokyo 108-8639, Japan*

Drug interactions which may cause harmful events are important for our health and new drug development. In the previous work, we extracted the drug interaction data from Japanese drug package inserts and generated the drug interaction network. The network contains a large number of drugs densely connected to each other, where drug targets and drug-metabolizing enzymes were shared in the drug interactions. In this study, we further analyzed the obtained drug interaction network by merging drugs into drug categories based on the Anatomical Therapeutic Chemical (ATC) classification. The merged data of drug interactions indicated drug properties that are related to drug interaction mechanisms or symptoms. We investigated the relationships between the drug groups and drug interaction mechanisms or symptoms.

Keywords: drug interactions; ATC classification system; KEGG.

1. Introduction

Combined use of multiple drugs may cause adverse events. Drug interactions can lead to an increase or a decrease of the drug effects or cause other serious reactions. For example, coadministration of a drug metabolized by Cytochrome P450 3A4 (CYP3A4) and the drug inhibiting CYP3A4, such as cyclosporine and clarithromycin, respectively, results in delayed clearance and elevated blood levels of the former drug, which increases and prolongs both the therapeutic and adverse effects [1]. Known information about potential risks of drug interactions is described in drug package inserts, which are documents attached with prescription medications to provide additional information about the drugs. The package insert information of Japanese marketed pharmaceutical products is provided by the Japan Pharmaceutical Information Center (JAPIC) database (http://database.japic.or.jp/nw/index), which is integrated with the KEGG DRUG database [2] to provide the GenomeNet pharmaceutical products database (http://www.genome.jp/kusuri/). Each JAPIC entry has its unique identifier (JAPIC ID) representing the package insert document of each marketed drug, which includes reported adverse drug interaction information. Adverse drug interactions are described with concrete names of compounds or names of drug classes. Interaction mechanisms and symptoms are classified according to the types of risks, contraindications or cautions for coadministration. On the contrary, the identifiers of the KEGG DRUG entries (D numbers) are associated with chemical compounds included in marketed drugs. In order to understand adverse drug interactions at the molecular perspective, we combined the compounds (D numbers) with the package inserts (JAPIC IDs) from many marketed products having the same main ingredients. In the previous work, we generated the drug interaction network with the extracted package insert information, and found the network including a number of drugs densely connected to each other, where drug targets and drug-metabolizing enzymes were shared with drug interactions [3]. The aim of this study is to characterize and classify drug interactions in order to suggest risks of potential drug interactions. We used the Anatomical Therapeutic Chemical (ATC) Classification System, which is controlled by the WHO Collaborating Centre for Drug Statistics Methodology, in order to merge the drugs having similar chemical or pharmacological properties to obtain the generic interpretation of the drug interaction network. Obtained data of drug groups, drug targets and drug-metabolizing enzymes were used to determine characteristics of drug interactions. Then we investigated the relationship between interaction characteristics and interaction mechanisms or symptoms.

2.   Method

2.1. *Datasets*

All drug interaction data were collected from the JAPIC database, containing the package insert data of 14,104 marketed drugs in Japan (as of June 2009) including the information on generic and trade names, physicochemical and pharmacokinetic properties, drug-metabolizing enzymes, and drug interactions. The drug interaction data is categorized as contraindications or cautions for coadministration, and is described using drug names or drug class names, interaction mechanisms and symptoms. The KEGG DRUG database (http://www.genome.jp/kegg/drug/) contains 8,912 approved drugs in either the U.S. or Japan as of July 2009. Each KEGG DRUG entry contains a D number (accession number), generic and trade names, molecular formula, chemical structure, target information, activity information, the ATC classification, and therapeutic category linked to the KEGG BRITE database (http://www.genome.jp/kegg/brite.html). The KEGG BRITE database is an ontology database and was used to retrieve information on functional hierarchies of drug targets, drug-metabolizing enzymes and compounds. We used the DrugBank database [4] to obtain more information on drug targets and drug-metabolizing enzymes.

2.2. *Drug groups based on the ATC classification*

We used the Anatomical Therapeutic Chemical (ATC) Classification System to group drugs. The ATC code, the identifier of the drug classification, consists of five different levels based to the organ or system on which the drugs act and/or the therapeutic and chemical characteristics. The first level represents anatomical main group, and consists of one letter (*e.g.*, "A" for alimentary tract and metabolism, "B" for blood and blood forming organs and "C" for cardiovascular system). The second and third levels represent therapeutic and pharmacological subgroups, respectively (*e.g.*, "C03" and "C03C" for diuretics and high-ceiling diuretics, respectively). The fourth and fifth levels indicate chemical subgroup and chemical substance, respectively (*e.g.*, "C03CA" and "C03CA01" sulfonamides and furosemide, respectively). In this study, the last three levels, *i.e.*, the third, fourth and fifth levels, were used to group drugs. For example, a drug erythromycin (D number: D00140) is classified into the ATC code "J01FA01" meaning a drug group "erythromycin", including eleven compounds such as erythromycin ethylsuccinate (D01361), erythromycin lactobionate (D02009) and erythromycin propionate (D02525). The fourth and the third levels of the code, J01FA and J01F, are defined as "macrolides" and "macrolides, lincosamides and streptogramins", respectively, and more compounds are included in those categories. Drugs with known drug interactions were grouped based on the ATC classification at each of those levels. There are cases where some D numbers share ATC codes (*e.g.*, clindamycin phosphate D01073 and clindamycin hydrochloride D02132 share three redundant codes "D10AF01", "G01AA10" and "J01FF01", all of which means "clindamycin"), which were manually merged into one drug group "clindamycin". The KEGG BRITE database was used to retrieve definitions of the drug groups. The drug groups were used to investigate the relationships with interaction mechanisms or symptoms at each ATC classification level.

2.3. *Drug interaction network*

Names of drugs or drug classes that interact with the other drugs were extracted from the drug interaction information in each JAPIC entry, and were replaced by the corresponding JAPIC IDs. The first type of drug interaction network was generated by representing marketed drugs (JAPIC IDs) as nodes and the interactions as edges. We progressively generated drug interaction networks at the different resolutions by merging the nodes according to different levels of the classification, *i.e.*, the JAPIC IDs, the D numbers and the drug groups at different ATC classification levels.

3. Results

We obtained 832 drug targets that were classified in 307 groups according to the definition in the KEGG BRITE database. The drug targets were associated with 926 drugs, among which 212 drugs were reported to act on biogenic amine (*e.g.*, adrenaline, dopamine, histamine, serotonin and acetylcholine) receptors. Table 1 lists major drug targets, definition of drug groups including drugs act on the targets and total number of the drugs.

Table 1. Drug targets and related drug groups

Drug target	Definition of drug group	# of drugs
Adrenaline receptors	Antipsychotics, Beta blocking agents	90
Cys4 zinc finger factors	Corticosteroids	74
Dopamine receptors	Antipsychotics, Dopaminergic agents	56
Histamine receptors	Antihistamines, Antipsychotics	54
Serotonin receptors	Antipsychotics, Antimigraine preparations	49
Oxidoreductases [EC:1.14.14.-]	Macrolides, Antifungals	44
Acetylcholine receptors	Antipsychotics, Anticholinergic agents	36

We searched the drug groups related to the drug-metabolizing enzymes. Table 2 shows examples of the drug group in which most drugs are metabolized by the same enzyme. For example, the drug group of glucocorticoids includes 30 drugs of which 29 are metabolized by the CYP3A4. Definition of the drug groups were retrieved from the KEGG BRITE database. We obtained 63 drug groups related to CYP3A, 34 drug groups to CYP2D6 and 14 drug groups to CYP1A2.

Table 2. Drug groups related to drug-metabolizing enzymes.

Drug-metabolizing enzyme	Definition of drug group	# of drugs
CYP3A4	Glucocorticoids	29 (of 30)
	Dihydropyridine Derivatives	10 (of 11)
	Protein kinase inhibitors	7 (of 7)
	Protease inhibitors	7 (of 7)
CYP2D6	Antidepressants	12 (of 15)
	Phenothiazines with piperazine structure	8 (of 9)
	Phenothiazines with aliphatic side-chain	5 (of 5)
CYP1A2	Natural and semisynthetic estrogens	5 (of 6)

We obtained drug interaction networks with different IDs and groups including JAPIC IDs, D numbers and the drug groups. Table 3 shows the number of edges and nodes in each drug interaction network.

Table 3. Number of nodes and edges in the drug interaction networks.

IDs / drug groups	# of edge	# of node
JAPIC ID	2201571	9227
D number	72611	2153
ATC at fifth level	37847	1617
ATC at fourth level	13637	539
ATC at third level	7067	211

The obtained data of drug targets and drug-metabolizing enzymes was attached to the drug interactions. Figure 1 shows examples of the drug interaction networks. Nodes represent D numbers (Fig. 1A) or drug groups at different ATC classification levels (Fig. 1B, C and D). Edges represent interactions between the nodes, and colored edges indicate the interactions by inhibition of CYPs (red edges) and additive effect or antagonism of biogenic amine receptors (green edges). Network size was reduced dependent on the level of drug groups based on the ATC classification, which helped to extract characteristics of the drug groups and interactions.

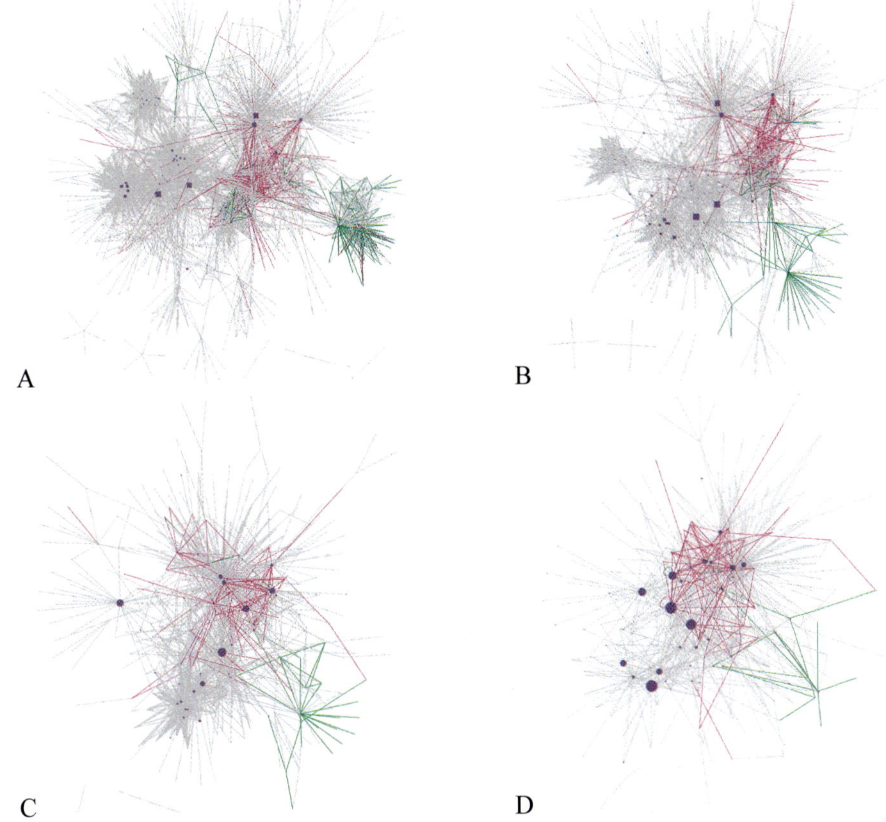

Figure 1. Drug interaction networks of contraindications for coadministration. In each network, nodes indicate as follows. A: D numbers. B: drug groups of "chemical substance" at fifth level of the ATC classification. C: drug groups of "chemical subgroup" at fourth level of the ATC classification. D: drug groups of "pharmacological subgroup" at third level of the ATC classification. Node size reflects number of interactions involved. Edges represent drug interactions between the two nodes. Green edges represent interactions which result from additive effects on the same biogenic amine receptor and red edges represent interactions caused by inhibiting CYP enzymes. The data used to generate these networks is limited to the interactions which were reported as contraindications for coadministration in order to decrease the nodes and edges in the networks.

In the interaction networks with the drug groups (Fig. 1B, C and D), nodes and edges shows drug properties and interaction characteristics. We searched interactions between the drug properties and interaction characteristics including interaction mechanisms and symptoms from the drug interaction networks at the third and fourth levels of the ATC classification. Figure 2 illustrates the extracted characteristic interactions in those networks. Large nodes and thick edges represent the obtained interactions including relationships between drug properties and interaction mechanisms or symptoms reported in the package insert information. In the network at the third level (Fig. 2A), we obtained 564 interactions between corticosteroids and adrenergics, 235 interactions between corticosteroids and insulins, 64 interactions of muscle relaxants with aminoglycosides or lincosamides and 67 interactions between beta-blocking agents and analgesics or antipyretics. At the fourth level (Fig. 2B), we obtained 92 interactions of fluoroquinolones or tetracyclines with calcium compounds, 55 interactions between angiotensin II antagonists and angiotensin-converting enzymes (ACE) inhibitors, 26 interactions between aminoglycosides and platinum compounds and 12 interactions between digoxin and angiotensin II antagonists.

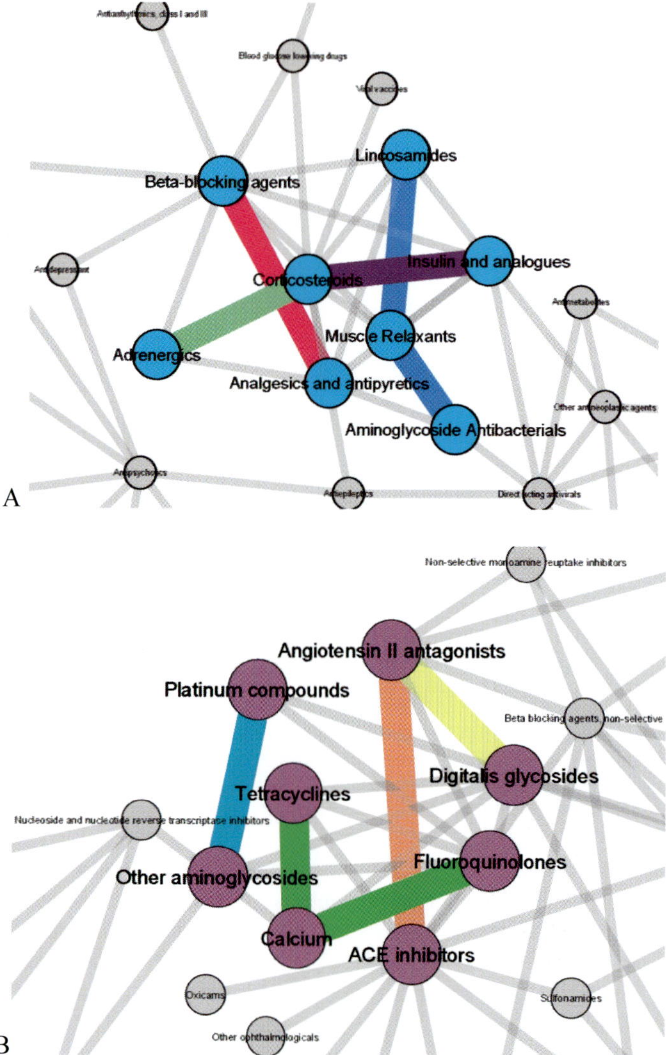

Figure 2. Characteristic interactions extracted from drug interaction networks at third (A) and fourth (B) levels of the ATC classification. The drug groups are illustrated as nodes with their definitions. Edges represent drug interactions between the drug groups. Large nodes and thick edges indicate the extracted interactions in which the drug properties related to the interaction mechanisms or symptoms. (A) The light green edge represents the interaction between corticosteroids and adrenergics. The purple edge indicates interaction between corticosteroids and insulins. Two blue edges show interactions of muscle relaxants with aminoglycosides or lincosamides. The red edge represents the interaction between beta-blocking agents and analgesics and

antipyretics. (B) Two green edges indicate interactions by administration of fluoroquinolones or tetracyclines with calcium compounds. The orange edge indicates the interaction between angiotensin II antagonists and angiotensin-converting enzymes. The light blue edge represents the interaction between aminoglycosides and platinum compounds. The yellow edge shows the interaction of digoxin with angiotensin II antagonists.

4. Discussion

We merged drugs into the drug groups according to the ATC classification to reduce data size and emphasize the relationships between drug properties and interaction characteristics including mechanisms and symptoms. Information on drug targets and drug-metabolizing enzymes indicate characteristics of the drug groups. We can use the drug target data to extract interactions between drugs acting on the same target, which lead to additive effects or antagonisms of the drug target. The drug-metabolizing enzyme data can also be used to search interactions related to the drug-metabolizing enzymes. The obtained data of drug groups related to CYPs indicate that an adverse drug interaction is caused by a combined use of the drug groups metabolized by a CYP and the drug groups that inhibit the same CYP.

We collected drug group pairs in which interaction mechanisms or symptoms are related to definitions of the drug groups manually. The interaction mechanisms and the symptoms were retrieved from drug package inserts and literature. The drugs in the adrenergic agent group acts to decrease serum potassium levels and the corticosteroid group enhances the action [5], which results in the increased effect of reduction in serum potassium level. The corticosteroids also decrease insulin sensitivity [6, 7], hence combined use of corticosteroids and insulin and analogues can cause decreased the insulin sensitivity and decreased antihyperglycemic effects. Non-steroidal anti-Inflammatory drugs (NSAIDs) are reported to inhibit prostaglandin synthesis [8-10]. That can cause elevated blood pressure and reduce the effect of antihypertensive drugs including beta-blocking agents [11]. Absorption of fluoroquinolones [12] and tetracyclines [13] are decreased by chelating effects with calcium compounds. ACE inhibitors [14] and angiotensin II antagonists [15, 16] lead to decreased blood pressure by different mechanisms. Thus the combined use of them results in increased antihypertensive effects. Aminoglycoside antibacterials [17-19] and lincosamides [20, 21] induce neuromuscular blockade, which leads to increased muscle relaxant effects. Both of aminoglycosides and platinum compounds can cause nephrotoxicity [22, 23] and thus coadmnistration of them increases the risk of the renal damage. Administration of digitalis glycosides with angiotensin II antagonists is reported to elevate serum digoxin level however the mechanism is unclear.

The interaction between ACE inhibitors and angiotensin II antagonists (Fig. 2B) is caused by their antihypertensive effects, however the two drugs act on different drug targets (ACE and AT1 receptors). In the KEGG PATHWAY database (http://www.genome.jp/kegg/pathway.html), the two drug targets are included in the same "renin-angiotensin system" pathway map (hsa04614). The map illustrates pathways of the angiotensin production from angiotensinogen and the action of angiotensin and includes related genes such as membrane metalloendopeptidase, neurolysin and chymase. Chymase converts angiotensin I to angiotensin II [24] as ACE, and drugs inhibiting chymase may cause the drug interactions with ACE inhibitors or angiotensin II antagonists, which enhance antihypertensive effects. In another instance, interactions between drugs acting on biogenic amine receptors and calcium channel blockers cause increased effects of antihypertensive, and the drug targets are illustrated in calcium signaling pathway (hsa04020). Hence, pathway information can help to search more characteristic interaction patterns with drug target data.

We searched characteristic drug interaction patterns from the networks. The extracted interaction patterns can be a clue to estimate unreported drug interactions and drugs interact with new pharmaceutical products.

Acknowledgments

This work was supported by the Ministry of Education, Culture, Sports, Science and Technology of Japan, and the Japan Science and Technology Agency. Computational resources were provided by the Bioinformatics Center and the Supercomputer Laboratory, Institute for Chemical Research, Kyoto University.

References

[1] Spicer, S. T., Liddle, C., Chapman, J. R., Barclay, P., Nankivell, B. J., Thomas, P., O'Connell, P. J., The mechanism of cyclosporin toxicity induced by clarythromycin. *Br. J. Clin. Pharmacol.* 43:194–196, 1997.

[2] Kanehisa, M., Araki, M., Goto, S., Hattori, M., Hirakawa, M., Itoh, M., Katayama, T., Kawashima, S., Okuda, S., Tokimatsu, T., Yamanishi, Y., KEGG for linking genomes to life and the environment. *Nucleic Acids Res.* 36:D480-D484, 2008.

[3] Takarabe, M., Okuda, S., Itoh, M., Tokimatsu, T., Goto, S., Kanehisa, M., Network analysis of adverse drug interactions. *Genome Inform.* 20, 252-259, 2008.

[4] Wishart, D. S., Knox, C., Guo, A. C., Cheng, D., Shrivastava, S., Tzur, D., Gautam, B., Hassanali, M. DrugBank: a knowledgebase for drugs, drug actions and drug targets. *Nucleic Acids Res.* 26:D901-D906, 2008.

[5] Taylor, D. R., Wilkins, G.T., Herbison, G. P., Flannery, E. M., Interaction between corticosteroid and beta-agonist drugs. Biochemical and cardiovascular effects in normal subjects. *Chest.* 102:519-524, 1992.

[6] Olefsky, J. M., Kimmerling, G., Effects of glucocorticoids on carbohydrate metabolism. *Am. J. Med. Sci.* 271:201-210, 1976.

[7] Pagano, G., Cavallo-Perin, P., Cassader, M., Bruno, A., Ozzello, A., Masciola, P., Dall'omo, A. M., Imbimbo, B., An in vivo and in vitro study of the mechanism of prednisone-induced insulin resistance in healthy subjects. *J. Clin. Invest.* 72:1814-1820, 1983.

[8] Vane, J. R., Inhibition of prostaglandin synthesis as a mechanism of action for aspirin-like drugs. *Nat. New Biol.* 231:232-235, 1971.

[9] Levine, L., Hinkle, P.M., Voelkel, E.F., Tashjian, A.H. Jr., Prostaglandin production by mouse fibrosarcoma cells in culture: inhibition by indomethacin and aspirin. *Biochem. Biophys. Res. Commun.* 1972, 47:888-896, 1972.

[10] Sykes, J. A., Maddox, I. S., Prostaglandin production by experimental tumours and its inhibition by non-steroidal-anti-inflammatory drugs. *Pol. J. Pharmacol. Pharm.* 26:83-91, 1974.

[11] Polónia, J., Interaction of antihypertensive drugs with anti-inflammatory drugs. *Cardiology* , 88:47-51, 1997.

[12] Polk, R. E., Drug interactions with ciprofloxacin and other fluoroquinolones. *Am. J. Med.* 87:76S-81S, 1989.

[13] Neuvonen, P. J., Interactions with the absorption of tetracyclines. *Drugs*, 11:45-54, 1976.

[14] Miller, E.D. Jr., Samuels, A.I., Haber, E., Barger, A.C., Inhibition of angiotensin conversion in experimental renovascular hypertension. *Science*, 177:1108-1109, 1972.

[15] Pals, D.T., Masucci, F.D., Sipos, F., Denning, G.S. Jr., A specific competitive antagonist of the vascular action of angiotensin. II. *Circ. Res.* 29:664-672, 1971.

[16] Pals, D.T., Masucci, F.D., Denning, G.S. Jr., Sipos, F., Fessler, D.C., Role of the pressor action of angiotensin II in experimental hypertension. *Circ. Res.* 29:673-681, 1971.

[17] Pittinger, C.B., Eryasa, Y., Adamson, R., Antibiotic-induced paralysis. *Anesth. Analg.* 49:487-501, 1970.

[18] Warner, W.A., Sanders, E., Neuromuscular blockade associated with gentamicin therapy. *J. Am. Med. Assoc.* 215:1153-1154, 1971.

[19] Dupuis, J.Y., Martin, R., Tétrault, J.P., Atracurium and vecuronium interaction with gentamicin and tobramycin. *Can. J. Anaesth.* 36:407-411, 1989.

[20] Daubeck, J.L., Daughety, M.J., Petty, C., Lincomycin-induced cardiac arrest: a case report and laboratory investigation. *Anesth. Analg.* 53:563-567, 1974.

[21] Samuelson, R.J., Giesecke, A.T. Jr., Kallus, F.T., Stanley, V.F., Lincomycin-curare interaction. *Anesth. Analg.* 54:103-105, 1975.

[22] Walker, R.J., Duggin, G.G., Drug nephrotoxicity. *Annu. Rev. Pharmacol. Toxicol.* 28:331-345, 1988.

[23] Madias, N.E., Harrington, J.T., Platinum nephrotoxicity. *Am. J. Med.* 65:307-314, 1978.

[24] Urata, H., Kinoshita, A., Misono, K.S., Bumpus, F.M., Husain, A., Identification of a highly specific chymase as the major angiotensin II-forming enzyme in the human heart. *J. Biol. Chem.* 265:22348-22357, 1990.

ANALYSIS AND PREDICTION OF NUTRITIONAL REQUIREMENTS USING STRUCTURAL PROPERTIES OF METABOLIC NETWORKS AND SUPPORT VECTOR MACHINES

TAKEYUKI TAMURA[1]

tamura@kuicr.kyoto-u.ac.jp

NILS CHRISTIAN[2]

nils.christian@mpimp-golm.mpg.de

KAZUHIRO TAKEMOTO[3]

takemoto@cb.k.u-tokyo.ac.jp

OLIVER EBENHÖH[4]

ebenhoeh@abdn.ac.uk

TATSUYA AKUTSU[1]

takutsu@kuicr.kyoto-u.ac.jp

[1] *Bioinformatics Center, Institute for Chemical Research, Kyoto University, Uji, Kyoto 611-0011, Japan*
[2] *Max Planck Institute of Molecular Plant Physiology, Potsdam-Golm, Germany*
[3] *Graduate School of Frontier Sciences, University of Tokyo, Kashiwanoha 5-1-5, Kashiwa, Chiba 277-8561, Japan*
[4] *Institute for Complex Systems and Mathematical Biology, University of Aberdeen, United Kingdom*

Properties of graph representation of genome scale metabolic networks have been extensively studied. However, the relationship between these structural properties and functional properties of the networks are still very unclear. In this paper, we focus on nutritional requirements of organisms as a functional property and study the relationship with structural properties of a graph representation of metabolic networks. In order to examine the relationship, we study to what extent the nutritional requirements can be predicted by using support vector machines from structural properties, which include degree exponent, edge density, clustering coefficient, degree centrality, closeness centrality, betweenness centrality and eigenvector centrality. Furthermore, we study which properties are influential to the nutritional requirements.

Keywords: metabolic networks; support vector machines; nutritional profile; centrality.

1. Introduction

Computational analysis of biological networks is becoming important in systems biology and bioinformatics. Among these networks, detailed and large-scale studies have mostly been performed on metabolic networks.[a] It may be due to the fact that rather accurate and large-scale network data are available from such databases as KEGG [11] and EcoCyc [12], compared to protein-protein interaction networks and gene regulatory networks.

---

[a]Extensive studies have been done on inference of protein-protein interaction networks and gene regulatory networks, but detailed and large-scale analysis of these networks are scarce.

In order to analyze structural properties of genome scale metabolic networks, many studies have been performed. In particular, such graph features as degree exponent, clustering coefficient, edge density, frequency of network motifs and various kinds of centrality measures have been extensively studied [3, 10, 21, 23]. However, the relationships between these structural features and functional properties of metabolic networks are still very unclear.

On the other hand, for prediction of some functional properties of metabolic networks, *flux balance analysis* (FBA) has been extensively studied [18, 20]. The FBA-based approach allows to infer an optimal flux distribution when the structure of a network and the target compounds whose production should be maximized are given. This approach has been successfully applied to predict flux distributions of *E.coli* [20] and to identify knock-out targets of enzymes/genes [5, 18]. Recently, as a complementary approach, Handorf et al. (2005) proposed the concept of *scope* [9]. The scope is the set of all possible metabolites obtained from a given set of *seed* compounds and a given structure of a metabolic network. Though the scope cannot do flux optimization, it is more tolerant against errors in the network and is much faster to calculate. Handorf et al. applied the scope to infer the minimal nutritional requirements that must be met to sustain maintenance or growth of an organism [8].

In this paper, we focus on *nutritional requirements* of organisms as a functional property and study their relationship with structural features. For this purpose we examine to what extent the nutritional requirements can be predicted by using *support vector machines* (SVMs) from structural features. As for global features of metabolic networks, we use average clustering coefficient, edge density, degree exponent, cyclic coefficient, subgraph concentration, assortativity coefficient and average path length. As for local features of metabolic networks (i.e., features for each node), we use degree centrality, closeness centrality, betweenness centrality and eigenvector centrality. As for training/test data, we use the nutritional profile generated by the scope-based method described in [8]. Furthermore, we study which structural features are influential to the nutritional requirements by examining combinations of structural features. The results show that use of either degree centrality or eigenvector centrality alone is effective for obtaining good prediction accuracy. The results also suggest that combination of centralities or combination of centrality and global features does not necessarily lead to improvement of prediction accuracy though global features are still useful for obtaining good prediction accuracy.

The organization of the paper is as follows. Section 2 and Section 3 explain the nutritional profiles and network features, respectively. Section 4 presents our prediction method using SVMs and global and local network features. Then, Section 5 provides the results of computational experiments. Finally, Section 6 discusses about the results and concludes with future work.

2. Generation of Nutritional Profiles

Each organism requires a minimal set of biochemical species, the nutrients, to allow for the production of metabolic precursors of higher molecules (e.g. proteins, RNA, DNA) and thus facilitate growth. For many organisms chemically defined nutrients are unknown. In [8] a method was introduced to computationally assess the minimal nutrients and a measure was derived that defines their essentiality, denoting how important these molecules are for the organism's growth.

Calculating nutritional profiles makes use of the method of network expansion [9], which calculates the set of producible metabolites from a given set of nutrients for a metabolic network. A biochemical reaction from the given metabolic network will operate if all its substrates are present. Subsequently, the reaction's products are added to the set of available metabolites. This procedure is iterated until no further metabolites are added, and the resulting set of metabolites is called the *scope* of the network for the given nutrients.

The method in [8] reverses this process: It calculates which nutrients are needed to produce a given set of target metabolites. Since the solution of this problem is not unique and highly combinatorial, a greedy algorithm is used to cover a large part of the solution space. Some molecules found in the solutions may be *replaceable* by others (e.g. because of nearly identical chemical composition). If a metabolite is replaceable by another, and the same holds true vice versa, these two metabolites are said to be *exchangeable*. For each organism one can therefore compile groups of so called *exchangeable resource metabolites*.

These organism specific groups are then distilled into global *resource types* by joining two metabolites if they are exchangeable in the majority of organism specific exchangeable resource metabolites. The relative occurrence of metabolites from a global resource type in the multiple nutrient sets for an organism defines the resource type's essentiality, reflected by a value in the interval $[0, 1]$. A nutrient profile for an organism is therefore formally defined as the vector of essentialities for the different resource types.

We have performed the above calculation for 447 organisms from the KEGG database [11], release 45. 45 resource types were defined. As targets we chose all metabolites that are present in at least 90% of the organisms' metabolic networks. In Fig. 1 we show the essentialities for a selected set of organisms to demonstrate

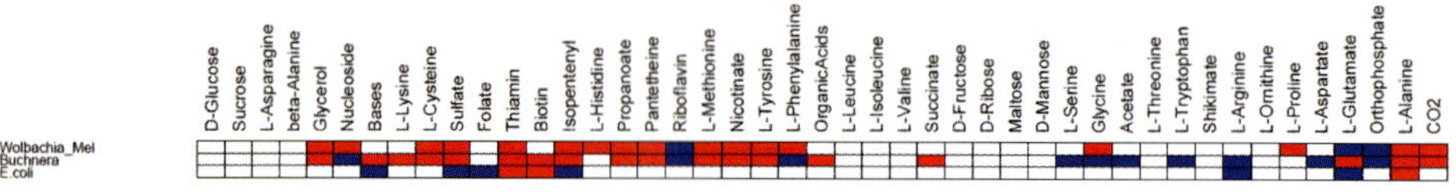

Fig. 1: Example of nutritional profiles.

White box: essentiality $= 0$, Blue box: $0 <$ essentiality < 0.9, Red box: essentiality ≥ 0.9

that the typical environment of an organism is reflected by the nutritional profiles. It can be clearly seen that the versatile *E. coli* needs a smaller set of nutrients to sustain growth compared to obligate symbionts and parasites such as *Buchnera* and *Wolbachia*, which satisfy their nutritional needs by using its host's metabolism.

The nutritional profiles thus describe a biological function, whose relationship to the structural properties of metabolic networks are studied in this work.

3. Structural Properties

To predict each nutrient profile from the network structure, we first constructed graphs from the metabolic networks, and then calculated nine global network parameters and four different centrality measures, where the centrality measures provide local features (i.e., each node's features) of a network.

3.1. *Dataset and Network Representation*

We downloaded the sets of metabolic reactions for 447 organisms from KEGG: Kyoto Encyclopedia of Genes and Genomes [11]. To emphasize the essential flow of metabolites, we neglected cofactor compounds of 71 types such as water and ATP in metabolic reactions.

The metabolic networks are represented by undirected graphs in which nodes and edges correspond to metabolites and substrate-product relationships, respectively. For example, considering a reaction S1+S2→P1+P2, metabolites S1 and S2 each connect to both products P1 and P2. That is, the edge list is as follows: (S1, P1), (S1, P2), (S2, P1), (S2, P2). In the case S1 and P1 are cofactor compounds the edge list is as follows: (S2, P2). Note that stoichiometric coefficients in the metabolic network are neglected.

To calculate network parameters (especially centrality measures), we extracted the largest connected component for each organism. In other words, small isolated clusters were removed.

3.2. *Global Network Parameters*

(i) The edge density: The edge density D is defined as the ratio of the number of edges E to the number of nodes N (i.e. $D = E/N$).

(ii) Clustering coefficient: The clustering coefficient is the average edge density at each node (i.e. at local levels), and is defined as $C = \sum_{i=1}^{N} c_i/N$. The value c_i is the local clustering coefficient defined as $2\Gamma_i/[k_i(k_i - 1)]$ [1], where Γ_i and k_i are the number of edges among neighbors of node i and the number of neighbors of node i, respectively.

(iii) Degree exponent: This characterizes the heterogeneity of network connectivity. In biological networks such as metabolic networks, the frequency of nodes with k edges $P(k)$ is well known to follow power-law distributions (reviewed in [3]): $P(k) \propto k^{-\gamma}$, where γ is the degree exponent. As the degree exponent increases, the probability that a node with large degree exists in a network decreases. That is, most nodes have similar degrees in the networks, indicating that the connectivity of the network is homogeneous. When the exponent becomes low, in contrast, the probability that a node with large degree exists in a network becomes high. That is, nodes tend to have different degrees in the networks, suggesting that the connectivity of the network is heterogeneous.

Assuming that the degree distribution of the metabolic networks follows a power law: $P(k) \propto k^{-\gamma}$, the degree exponent γ is extracted using maximum likelihood estimate given by the formula $\gamma = 1 + N \left[\sum_{i=1}^{N} \ln(k_i/k_{min}) \right]^{-1}$ [16], where k_{min} is the smallest degree (the number of neighbors) in the network.

(iv) Cyclic coefficient The cyclic coefficient [13] is an extended clustering coefficient. The clustering coefficient only characterizes connections among neighbors (i.e. triangles or cycles of length 3), however, the cyclic coefficient can detect cycles of length more than 3 in addition to triangles. The cyclic coefficient is defined as $R = \sum_{i=1}^{N} r_i/N$, where r_i is $2 \sum_{\langle jh \rangle} (L_{jh}^{i} - 2)^{-1}/[k_i(k_i - 1)]$. $\langle jh \rangle$ denotes all pairs of neighbors of node i, and L_{jh}^{i} is the length of the smallest cycle that passes through node i and its two neighbors j and h.

(v–vii) Subgraph concentrations: The (nt)-subgraph consists of a central node, $n - 1$ neighbors and $n - 1 + t$ edges, where t denotes the number of edges among the neighbors [22]. That is, a subgraph composed of n nodes contains $(n-1)(n-2)/2+1$ different subgraphs because the maximal value of t is $\binom{n-1}{2}$.

The subgraph concentration [22] denotes a fraction of (nt)-subgraph abundance in all types of n-node subgraphs, and is defined as $S_{nt} = s_{nt}/ \sum_{i=0}^{(n-1)(n-2)/2} s_{ni}$, where s_{nt} corresponds to (nt)-subgraph abundance.

In this paper, we focus on (31)-subgraphs (i.e. triangles), (42)-subgraphs (i.e. squares including two triangles), and (43)-subgraphs (i.e. 4-node complete graphs).

(viii) Assortative coefficient: The assortative coefficient [15] can be thought of as a compendium parameter of the correlation coefficient between the degree (the number of neighbors) of a node and the degrees of neighbors, and is defined as $r = (4\langle k_i k_j \rangle - \langle k_i + k_j \rangle^2)/(2\langle k_i^2 + k_j^2 \rangle - \langle k_i + k_j \rangle^2)$, where k_i and k_j are the degrees of two nodes at the ends of an edge, and $\langle \cdots \rangle$ denotes the average over all edges.

(ix) Average path length: The average path length is the average length of the shortest paths between two nodes, and is defined as $L = \sum_{i,j} d_{ij}/[N(N-1)]$ [1], where d_{ij} is the shortest path length between nodes i and j.

3.3. *Centrality Measures*

(I) Degree centrality: Assuming correlation between the centrality (or importance) of a node and the degree (the number of neighbors) of the node, the degree centrality of node i is defined as $C_D(i) = k_i/(N-1)$ [7], where k_i is the degree of node i.

(II) Closeness centrality: When the average path length between a node and all other nodes is relatively short, the centrality of such a node can be considered high.

Therefore the closeness centrality of node i is defined as $C_C(i) = [\sum_{j=1, j \neq i}^{N} d_{ij}]^{-1}$ [7].
(III) Betweenness centrality: If a walker moves from one node to another via their shortest path, then well-passed nodes are defined to have a high centrality. Based on this, the betweenness centrality of node i is defined as $C_B(i) = \sum_{s \neq t \neq i} \sigma_{st}(i)/\sigma_{st}$ [7], where $\sigma_{st}(i)$ and σ_{st} are the number of shortest paths between nodes s and t on which there is node i and the number of shortest paths between nodes s and t, respectively. For normalization, the betweenness centrality is finally divided by the maximum value.
(IV) Eigenvector centrality: This is a higher version of the degree centrality. The degree centrality is only based on the number of neighbors. However, the eigenvector centrality can consider neighbors' centralities. The centrality $C_E(i)$ of node i is proportional to the average of the centralities of neighbors of node i: $C_E(i) = \lambda^{-1} \sum_{j=1}^{N} M_{ij} \cdot C_E(j)$, where λ is a constant and M_{ij} is the adjacency matrix. $M_{ij} = 1$ if node i connects to node j, and $M_{ij} = 0$ otherwise. With $\boldsymbol{x} = (C_E(1), \ldots, C_E(N))$ this equation can be rewritten as $\lambda \boldsymbol{x} = \boldsymbol{M} \cdot \boldsymbol{x}$. The eigenvector centrality is defined as the eigenvector with the largest eigenvalue [4].

4. Prediction by Support Vector Machines

SVM is a kind of statistical learning method and is basically used for binary classification. Let POS and NEG be the sets of positive examples and negative examples in a training data set, where each example is represented as a point in d-dimensional Euclidean space, and the corresponding d-dimensional vector is called a *feature vector*. Then, an SVM finds a hyperplane h such that the distance between h and the closest point is the maximum (i.e., the margin is maximized) under the condition that all points in POS lie above h, and all points in NEG lie below h. Once this h is obtained, we can infer that a new test data is positive (resp. negative) if it lies above h (resp. below h). If it is impossible to completely separate positive points from negative points, the soft margin (weighted combination of the margin and classification errors) is optimized.

SVM does not usually use feature vectors directly. It uses feature vectors in the form of kernel functions, where the kernel function is basically defined as the inner product between two feature vectors. Let $\mathbf{v}_{i_1}$ and $\mathbf{v}_{i_2}$ be the feature vectors of examples i_1 and i_2, respectively. Then, the value of the kernel function $K(\mathbf{v}_{i_1}, \mathbf{v}_{i_2})$ for this pair is calculated by $\mathbf{v}_{i_1} \cdot \mathbf{v}_{i_2}$. In some cases, more complex kernel functions are used to obtain better prediction performance. For example, $K(\mathbf{v}_{i_1}, \mathbf{v}_{i_2}) = exp(-\gamma \cdot ||\mathbf{v}_{i_1} - \mathbf{v}_{i_2}||^2)$ is frequently used. This kernel can be interpreted as the inner product between two infinite-dimensional vectors $\phi(\mathbf{v}_{i_1})$ and $\phi(\mathbf{v}_{i_2})$, where $\phi(\mathbf{v}_{i_j})$ is obtained from $\mathbf{v}_{i_j}$ (i.e., feature vector $\mathbf{v}_{i_j}$ is transformed into the infinite-dimensional feature vector $\phi(\mathbf{v}_{i_j})$). It is to be noted that we cannot show the exact form of $\phi(\mathbf{v}_{i_j})$ or calculate $\phi(\mathbf{v}_{i_j})$ explicitly, but can show the existence of $\phi(\mathbf{v}_{i_j})$ and can calculate $K(\phi(\mathbf{v}_{i_1}), \phi(\mathbf{v}_{i_2}))$ efficiently. This property (i.e., kernel functions can be efficiently computed without explicitly computing feature vectors) is known as *kernel trick*. For details of SVMs and kernel functions, see [6, 19].

Let m and n be the numbers of compounds and organisms respectively ($m = 2140$, $n = 447$). Three types of matrices A, B, C appear in this prediction problem. As for A, $a_{i,j}$ represents the essentiality of the j-th resource type for the i-th organism. As for B, $b_{i,j}$ represents the value of the j-th global feature of the i-th organism. As for C, $c_{i,j,k}$ represents the value of the j-th local feature of the i-th compound and k-th organism. The purpose of the problem is to predict $a_{i,j}$ when $b_{i,j}$ and $c_{i,j,k}$ are given. $a_{i,j}$ is converted to a binary matrix with a threshold of 0.9. That is, if $a_{i,j} \geq 0.9$, it is treated as 1, otherwise it is treated as 0.

By shuffling organisms of A into five groups, we conducted fivefold cross validation test where information of B and C was used as the feature vector.

Each organism has at most $9+4m$ entries in the feature vector, where 9 elements are based on the information of B and $4m$ elements are based on the information from C. The kernel value for organisms i_1 and i_2 with the feature vectors $\mathbf{v}_{i_1}$ and $\mathbf{v}_{i_2}$ is calculated by $K(\mathbf{v}_{i_1}, \mathbf{v}_{i_2}) = exp(-10 \cdot ||\mathbf{v}_{i_1} - \mathbf{v}_{i_2}||^2)$.

In order to implement SVM, we used the software GIST [17]. We evaluated the prediction performance of our method by calculating sensitivity (sen), specificity (spe), negative sensitivity (senn), negative specificity (spen) and Matthew's correlation coefficient (MCC) [14] for each resource type. The definitions of these measures are as follows: Sensitivity=$\frac{TP}{TP+FN}$, Specificity=$\frac{TP}{TP+FP}$, Negative Sensitivity=$\frac{TN}{TN+FP}$, Negative Specificity=$\frac{TN}{TN+FN}$.

$$MCC(l) = \frac{TP \cdot TN - FP \cdot FN}{\sqrt{(TP + FN)(TP + FP)(TN + FP)(TN + FN)}}.$$

TP, TN, FP, FN denote true positive, true negative, false positive and false negative respectively.

5. Results

Although we conducted the computer experiments for 45 resource types, we extracted 17 resource types which satisfy $1/3 \leq (TP + FN)/(TN + FP) \leq 3$ for the following tables. This is reasonable since a fair evaluation of predictive accuracy is difficult if the number of positive examples or negative examples is too small.

Prediction results of fivefold cross validation where the feature vector consists of the 9 global parameters explained in Section 3.2 are shown in Table 1. Average of sensitivity, specificity, negative-sensitivity and negative specificity are 0.7199, 0.7344, 0.7261 and 0.7414 respectively. Since all these values are larger than 0.7, we can infer that information of global structural properties of metabolic networks is useful for predicting nutritional requirements of organisms.

However, Tables 2-4 show that some local properties of networks are more useful to predict nutritional requirements than global properties. Table 2 shows prediction results of fivefold cross validation where only degree centrality is used to calculate the feature vector, that is, the number of elements of the feature vector is m. Average

Table 1: Prediction result of fivefold cross validation where the feature vector consists of 9 global parameters explained in Section 3.1.

Resource type	FP	FN	TP	TN	sen	spe	senn	spen	mcc
Bases	63	94	96	194	0.5052	0.6037	0.7548	0.6736	0.2686
CO2	54	60	159	174	0.7260	0.7464	0.7631	0.7435	0.4896
D.Glucose	64	57	167	159	0.7455	0.7229	0.7130	0.7361	0.4587
Folate	72	36	274	65	0.8838	0.7919	0.4744	0.6435	0.3950
Glycerol	58	52	118	219	0.6941	0.6704	0.7906	0.8081	0.4816
L.Alanine	29	38	140	240	0.7865	0.8284	0.8921	0.8633	0.6851
L.Cysteine	62	87	92	206	0.5139	0.5974	0.7686	0.7030	0.2914
L.Lysine	62	45	175	165	0.7954	0.7383	0.7268	0.7857	0.5232
L.Methionine	67	48	245	87	0.8361	0.7852	0.5649	0.6444	0.4151
L.Phenylalanine	30	62	72	283	0.5373	0.7058	0.9041	0.8202	0.4819
L.Tyrosine	26	54	77	290	0.5877	0.7475	0.9177	0.8430	0.5463
Nicotinate	42	65	152	188	0.7004	0.7835	0.8173	0.7430	0.5222
Nucleoside	71	67	170	139	0.7173	0.7053	0.6619	0.6747	0.3796
OrganicAcids	38	48	77	284	0.6160	0.6695	0.8819	0.8554	0.5113
Pantetheine	54	43	253	97	0.8547	0.8241	0.6423	0.6928	0.5069
Riboflavin	77	47	236	87	0.8339	0.7539	0.5304	0.6492	0.3833
Sulfate	65	29	277	76	0.9052	0.8099	0.5390	0.7238	0.4869
Average					0.7199	0.7344	0.7261	0.7414	0.4604

of sensitivity, specificity, negative-sensitivity and negative specificity are 0.7232, 0.7922, 0.7649 and 0.7818, respectively. Since these values are larger than those of Table 1 by 0.0033, 0.0578, 0.0388 and 0.0404, it follows that degree centrality is more useful than global properties.

Similarly, Table 3 corresponds to the case where only closeness centrality is taken into consideration. However, specificity and negative sensitivity (0.5310 and 0.1816) are less than those of Table 1 by 0.2034 and 0.5445, respectively, although sensitivity and negative-specificity (0.9830 and 0.9090) are larger than those of Table 1 by 0.2631 and 0.1676. Since mcc of 0.2666 is less than mcc of Table 1 (0.4604) by 0.1938, we can conclude that closeness centrality is not useful when compared to global properties.

Finally, Table 4 corresponds to the case where only eigenvector centrality is used to calculate the feature vector. Although sensitivity (0.6907) is less than that of Table 1 by 0.0292, specificity, negative sensitivity and negative specificity (0.8431, 0.8179, 0.7774) are larger than those of Table 1 by 0.1087, 0.0918 and 0.0360, respectively. Since mcc of 0.5583 is larger than that of Table 1 (0.4604) by 0.0979 eigenvector centrality is more useful than global properties.

As said above, using only degree centrality or eigenvector centrality yields better predictive accuracy than using only global properties, but closeness centrality turns out not to be useful. The SVM calculations for betweenness centrality did not finish within three days and were aborted.

We also combined global and local properties to calculate the feature vector. Tables 5-9 show that combining global and local properties appropriately may yield

Table 2: Prediction result of fivefold cross validation where only degree centrality is used to compute the feature vector.

Resource type	FP	FN	TP	TN	sen	spe	senn	spen	mcc
Bases	54	86	104	203	0.5473	0.6582	0.7898	0.7024	0.3487
CO2	32	59	160	196	0.7305	0.8333	0.8596	0.7686	0.5960
D.Glucose	65	48	176	158	0.7857	0.7302	0.7085	0.7669	0.4957
Folate	68	17	293	69	0.9451	0.8116	0.5036	0.8023	0.5249
Glycerol	30	66	104	247	0.6117	0.7761	0.8916	0.7891	0.5334
L.Alanine	20	35	143	249	0.8033	0.8773	0.9256	0.8767	0.7414
L.Cysteine	40	66	113	228	0.6312	0.7385	0.8507	0.7755	0.4977
L.Lysine	79	36	184	148	0.8363	0.6996	0.6519	0.8043	0.4960
L.Methionine	53	33	260	101	0.8873	0.8306	0.6558	0.7537	0.5634
L.Phenylalanine	11	76	58	302	0.4328	0.8405	0.9648	0.7989	0.5043
L.Tyrosine	12	72	59	304	0.4503	0.8309	0.9620	0.8085	0.5135
Nicotinate	12	79	138	218	0.6359	0.9200	0.9478	0.7340	0.6178
Nucleoside	79	43	194	131	0.8185	0.7106	0.6238	0.7528	0.4528
OrganicAcids	15	67	58	307	0.4640	0.7945	0.9534	0.8208	0.5068
Pantetheine	58	41	255	93	0.8614	0.8146	0.6158	0.6940	0.4928
Riboflavin	64	34	249	100	0.8798	0.7955	0.6097	0.7462	0.5150
Sulfate	72	8	298	69	0.9738	0.8054	0.4893	0.8961	0.5700
Average					0.7232	0.7922	0.7649	0.7818	0.5277

Table 3: Prediction result of fivefold cross validation where only closeness centrality is used to calculate the feature vector.

Resource type	FP	FN	TP	TN	sen	spe	senn	spen	mcc
Bases	203	5	185	54	0.9736	0.4768	0.2101	0.9152	0.2684
CO2	196	0	219	32	1.0000	0.5277	0.1403	1.0000	0.2721
D.Glucose	180	3	221	43	0.9866	0.5511	0.1928	0.9347	0.2952
Folate	101	4	306	36	0.9870	0.7518	0.2627	0.9000	0.4035
Glycerol	216	2	168	61	0.9882	0.4375	0.2202	0.9682	0.2908
L.Alanine	240	1	177	29	0.9943	0.4244	0.1078	0.9666	0.1999
L.Cysteine	224	4	175	44	0.9776	0.4385	0.1641	0.9166	0.2244
L.Lysine	172	5	215	55	0.9772	0.5555	0.2422	0.9166	0.3219
L.Methionine	134	5	288	20	0.9829	0.6824	0.1298	0.8000	0.2332
L.Phenylalanine	259	2	132	54	0.9850	0.3375	0.1725	0.9642	0.2181
L.Tyrosine	259	0	131	57	1.0000	0.3358	0.1803	1.0000	0.2461
Nicotinate	206	2	215	24	0.9907	0.5106	0.1043	0.9230	0.2031
Nucleoside	165	9	228	45	0.9620	0.5801	0.2142	0.8333	0.2700
OrganicAcids	266	4	121	56	0.9680	0.3126	0.1739	0.9333	0.1868
Pantetheine	129	6	290	22	0.9797	0.6921	0.1456	0.7857	0.2448
Riboflavin	130	8	275	34	0.9717	0.6790	0.2073	0.8095	0.2957
Sulfate	110	4	302	31	0.9869	0.7330	0.2198	0.8857	0.3576
Average					0.9830	0.5310	0.1816	0.9090	0.2666

better accuracies than using only either global or local properties.

Tables 5-9 correspond to the cases where degree centrality, closeness centrality, betweenness centrality, eigenvector centrality alone and all these four properties together are used to calculate the feature vectors in addition to the 9 global features.

Table 4: Prediction result of fivefold cross validation where only eigenvector centrality is used to calculate the feature vector.

Resource type	FP	FN	TP	TN	sen	spe	senn	spen	mcc
Bases	40	77	113	217	0.5947	0.7385	0.8443	0.7380	0.4574
CO2	22	60	159	206	0.7260	0.8784	0.9035	0.7744	0.6411
D.Glucose	57	28	196	166	0.8750	0.7747	0.7443	0.8556	0.6248
Folate	69	16	294	68	0.9483	0.8099	0.4963	0.8095	0.5248
Glycerol	18	85	85	259	0.5000	0.8252	0.9350	0.7529	0.5015
L.Alanine	10	59	119	259	0.6685	0.9224	0.9628	0.8144	0.6821
L.Cysteine	17	86	93	251	0.5195	0.8454	0.9365	0.7448	0.5188
L.Lysine	47	49	171	180	0.7772	0.7844	0.7929	0.7860	0.5703
L.Methionine	37	50	243	117	0.8293	0.8678	0.7597	0.7005	0.5786
L.Phenylalanine	4	79	55	309	0.4104	0.9322	0.9872	0.7963	0.5382
L.Tyrosine	4	79	52	312	0.3969	0.9285	0.9873	0.7979	0.5283
Nicotinate	7	86	131	223	0.6036	0.9492	0.9695	0.7216	0.6201
Nucleoside	73	41	196	137	0.8270	0.7286	0.6523	0.7696	0.4887
OrganicAcids	11	72	53	311	0.4240	0.8281	0.9658	0.8120	0.4995
Pantetheine	33	56	240	118	0.8108	0.8791	0.7814	0.6781	0.5745
Riboflavin	44	43	240	120	0.8480	0.8450	0.7317	0.7361	0.5805
Sulfate	77	5	301	64	0.9836	0.7962	0.4539	0.9275	0.5627
Average					0.6907	0.8431	0.8179	0.7774	0.5583

Table 5: Prediction result of fivefold cross validation where the feature vector is calculated by 9 global parameters and degree centrality. (The total number of elements of the feature vector is $9+m$.)

Resource type	FP	FN	TP	TN	sen	spe	senn	spen	mcc
Bases	42	90	100	215	0.5263	0.7042	0.8365	0.7049	0.3853
CO2	34	51	168	194	0.7671	0.8316	0.8508	0.7918	0.6207
D.Glucose	72	38	186	151	0.8303	0.7209	0.6771	0.7989	0.5136
Folate	77	12	298	60	0.9612	0.7946	0.4379	0.8333	0.5007
Glycerol	30	56	114	247	0.6705	0.7916	0.8916	0.8151	0.5841
L.Alanine	16	43	135	253	0.7584	0.8940	0.9405	0.8547	0.7234
L.Cysteine	32	71	108	236	0.6033	0.7714	0.8805	0.7687	0.5112
L.Lysine	81	36	184	146	0.8363	0.6943	0.6431	0.8021	0.4879
L.Methionine	59	32	261	95	0.8907	0.8156	0.6168	0.7480	0.5349
L.Phenylalanine	11	80	54	302	0.4029	0.8307	0.9648	0.7905	0.4780
L.Tyrosine	10	73	58	306	0.4427	0.8529	0.9683	0.8073	0.5210
Nicotinate	9	75	142	221	0.6543	0.9403	0.9608	0.7466	0.6501
Nucleoside	66	43	194	144	0.8185	0.7461	0.6857	0.7700	0.5102
OrganicAcids	17	64	61	305	0.4880	0.7820	0.9472	0.8265	0.5146
Pantetheine	56	35	261	95	0.8817	0.8233	0.6291	0.7307	0.5320
Riboflavin	58	29	254	106	0.8975	0.8141	0.6463	0.7851	0.5709
Sulfate	73	6	300	68	0.9803	0.8042	0.4822	0.9189	0.5784
Average					0.7300	0.8007	0.7682	0.7937	0.5422

In Table 5, sensitivity, specificity, negative sensitivity, negative specificity and mcc were 0.7300, 0.8007, 0.7682, 0.7937 and 0.5422 and they are larger than those of Table 2 by 0.0068, 0.0078, 0.0033, 0.0119 and 0.0145. Therefore, we can conclude

Table 6: Prediction result of fivefold cross validation where the feature vector is calculated by 9 global parameters and closeness centrality.

Resource type	FP	FN	TP	TN	sen	spe	senn	spen	mcc
Bases	203	5	185	54	0.9736	0.4768	0.2101	0.9152	0.2684
CO2	196	0	219	32	1.0000	0.5277	0.1403	1.0000	0.2721
D.Glucose	180	3	221	43	0.9866	0.5511	0.1928	0.9347	0.2952
Folate	101	4	306	36	0.9870	0.7518	0.2627	0.9000	0.4035
Glycerol	216	2	168	61	0.9882	0.4375	0.2202	0.9682	0.2908
L.Alanine	239	1	177	30	0.9943	0.4254	0.1115	0.9677	0.2040
L.Cysteine	224	4	175	44	0.9776	0.4385	0.1641	0.9166	0.2244
L.Lysine	172	5	215	55	0.9772	0.5555	0.2422	0.9166	0.3219
L.Methionine	134	5	288	20	0.9829	0.6824	0.1298	0.8000	0.2332
L.Phenylalanine	259	2	132	54	0.9850	0.3375	0.1725	0.9642	0.2181
L.Tyrosine	259	0	131	57	1.0000	0.3358	0.1803	1.0000	0.2461
Nicotinate	206	2	215	24	0.9907	0.5106	0.1043	0.9230	0.2031
Nucleoside	165	9	228	45	0.9620	0.5801	0.2142	0.8333	0.2700
OrganicAcids	266	4	121	56	0.9680	0.3126	0.1739	0.9333	0.1868
Pantetheine	129	6	290	22	0.9797	0.6921	0.1456	0.7857	0.2448
Riboflavin	130	8	275	34	0.9717	0.6790	0.2073	0.8095	0.2957
Sulfate	110	4	302	31	0.9869	0.7330	0.2198	0.8857	0.3576
Average					0.9830	0.5310	0.1819	0.9090	0.2668

Table 7: Prediction result of fivefold cross validation where the feature vector is calculated by 9 global parameters and betweenness centrality.

Resource type	FP	FN	TP	TN	sen	spe	senn	spen	mcc
Bases	68	87	103	189	0.5421	0.6023	0.7354	0.6847	0.2822
CO2	45	52	167	183	0.7625	0.7877	0.8026	0.7787	0.5658
D.Glucose	77	57	167	146	0.7455	0.6844	0.6547	0.7192	0.4019
Folate	76	21	289	61	0.9322	0.7917	0.4452	0.7439	0.4496
Glycerol	44	53	117	233	0.6882	0.7267	0.8411	0.8146	0.5353
L.Alanine	24	33	145	245	0.8146	0.8579	0.9107	0.8812	0.7323
L.Cysteine	48	74	105	220	0.5865	0.6862	0.8208	0.7482	0.4208
L.Lysine	67	44	176	160	0.8000	0.7242	0.7048	0.7843	0.5067
L.Methionine	64	38	255	90	0.8703	0.7993	0.5844	0.7031	0.4780
L.Phenylalanine	19	72	62	294	0.4626	0.7654	0.9392	0.8032	0.4781
L.Tyrosine	14	64	67	302	0.5114	0.8271	0.9556	0.8251	0.5520
Nicotinate	31	67	150	199	0.6912	0.8287	0.8652	0.7481	0.5665
Nucleoside	65	57	180	145	0.7594	0.7346	0.6904	0.7178	0.4512
OrganicAcids	27	53	72	295	0.5760	0.7272	0.9161	0.8477	0.5319
Pantetheine	53	39	257	98	0.8682	0.8290	0.6490	0.7153	0.5306
Riboflavin	67	34	249	97	0.8798	0.7879	0.5914	0.7404	0.4990
Sulfate	72	13	293	69	0.9575	0.8027	0.4893	0.8414	0.5365
Average					0.7322	0.7625	0.7409	0.7704	0.5011

that combining degree centrality and global properties yields better predictive accuracies than using only degree centrality.

In Table 6 mcc is 0.2668, thus taking closeness centrality into consideration still yields poor accuracies even when global properties are also taken into account.

Table 8: Prediction result of fivefold cross validation where the feature vector is calculated by 9 global parameters and eigenvector centrality.

Resource type	FP	FN	TP	TN	sen	spe	senn	spen	mcc
Bases	51	79	111	206	0.5842	0.6851	0.8015	0.7228	0.3967
CO2	28	59	160	200	0.7305	0.8510	0.8771	0.7722	0.6154
D.Glucose	65	30	194	158	0.8660	0.7490	0.7085	0.8404	0.5819
Folate	64	22	288	73	0.9290	0.8181	0.5328	0.7684	0.5205
Glycerol	19	77	93	258	0.5470	0.8303	0.9314	0.7701	0.5360
L.Alanine	12	50	128	257	0.7191	0.9142	0.9553	0.8371	0.7119
L.Cysteine	22	82	97	246	0.5418	0.8151	0.9179	0.7500	0.5097
L.Lysine	50	49	171	177	0.7772	0.7737	0.7797	0.7831	0.5569
L.Methionine	34	50	243	120	0.8293	0.8772	0.7792	0.7058	0.5957
L.Phenylalanine	11	76	58	302	0.4328	0.8405	0.9648	0.7989	0.5043
L.Tyrosine	10	76	55	306	0.4198	0.8461	0.9683	0.8010	0.5012
Nicotinate	6	80	137	224	0.6313	0.9580	0.9739	0.7368	0.6485
Nucleoside	73	44	193	137	0.8143	0.7255	0.6523	0.7569	0.4745
OrganicAcids	17	67	58	305	0.4640	0.7733	0.9472	0.8198	0.4939
Pantetheine	36	53	243	115	0.8209	0.8709	0.7615	0.6845	0.5688
Riboflavin	42	44	239	122	0.8445	0.8505	0.7439	0.7349	0.5869
Sulfate	69	5	301	72	0.9836	0.8135	0.5106	0.9350	0.6082
Average					0.7021	0.8231	0.8121	0.7775	0.5536

Table 9: Prediction result of fivefold cross validation where the feature vector is calculated by 9 global parameters, degree centrality, closeness centrality, betweenness centrality and eigenvector centrality. (The total number of elements of the feature vector is $9+4m$.)

Resource type	FP	FN	TP	TN	sen	spe	senn	spen	mcc
Bases	204	5	185	53	0.9736	0.4755	0.2062	0.9137	0.2646
CO2	201	0	219	27	1.0000	0.5214	0.1184	1.0000	0.2484
D.Glucose	185	3	221	38	0.9866	0.5443	0.1704	0.9268	0.2719
Folate	106	4	306	31	0.9870	0.7427	0.2262	0.8857	0.3661
Glycerol	218	1	169	59	0.9941	0.4366	0.2129	0.9833	0.2949
L.Alanine	241	1	177	28	0.9943	0.4234	0.1040	0.9655	0.1957
L.Cysteine	228	4	175	40	0.9776	0.4342	0.1492	0.9090	0.2087
L.Lysine	177	4	216	50	0.9818	0.5496	0.2202	0.9259	0.3099
L.Methionine	134	5	288	20	0.9829	0.6824	0.1298	0.8000	0.2332
L.Phenylalanine	264	2	132	49	0.9850	0.3333	0.1565	0.9607	0.2040
L.Tyrosine	264	0	131	52	1.0000	0.3316	0.1645	1.0000	0.2336
Nicotinate	206	1	216	24	0.9953	0.5118	0.1043	0.9600	0.2169
Nucleoside	167	8	229	43	0.9662	0.5782	0.2047	0.8431	0.2684
OrganicAcids	269	3	122	53	0.9760	0.3120	0.1645	0.9464	0.1906
Pantetheine	129	5	291	22	0.9831	0.6928	0.1456	0.8148	0.2557
Riboflavin	130	7	276	34	0.9752	0.6798	0.2073	0.8292	0.3048
Sulfate	116	4	302	25	0.9869	0.7224	0.1773	0.8620	0.3098
Average					0.9850	0.5278	0.1684	0.9133	0.2575

In Table 7, sensitivity, specificity, negative sensitivity and negative specificity were 0.7322, 0.7625, 0.7409 and 0.7704, respectively. Since all these values are larger than

0.7 and mcc is also larger than 0.5, we can conclude that combining betweenness centrality and global properties yields good predictive accuracies. In Table 8, although sensitivity and negative specificity (0.7021 and 0.7775) are larger than those of Table 4 by 0.0114 and 0.0001, specificity and negative-sensitivity (0.8231 and 0.8121) are smaller than those of Table 4 by 0.0200 and 0.0058. Since mcc of 0.5536 is smaller than that of Table 4 by 0.0047, it can be said that combining global properties with eigenvector centrality failed to improve the predictive accuracy. Finally, since mcc of Table 9 is 0.2575, we can conclude that combining all local and global features did not succeed in improving predictive accuracies.

6. Discussion and Conclusion

Here, we discuss the results of our computational experiments. First, we speculate about reasons for the predictability of nutrient profiles using network parameters. It is expected that metabolic pathways around compounds corresponding to nutrients are less dense because such compounds are not synthesized due to exogenous supply. In addition, the nutrient compounds are probably located at the periphery of metabolic networks for similar reasons. Thus, the global network parameters, which characterize network density, and the centralities of each node might be useful to predict nutrient profiles.

Overall, we could predict nutrient profiles from the network structure. Using the closeness centrality, however, predictions for several nutrient profiles showed relatively low accuracy. This might be caused by small-worldness of metabolic networks [23] that are represented as substrate-product relationships. As our metabolic networks are constructed by this representation, they have small-world features, indicating that all nodes are linked by short paths. Since the closeness centrality is based on the average path length as above, it is roughly homogenous among nodes due to the small-worldness. For this reason, the prediction using the closeness centrality was not effective. To predict nutrient profiles more accurately using network parameters, we might need to consider more appropriate network representations. For example, metabolic networks are not small-world when they are defined by atomic mappings [2] instead of substrate-product relationships. Accordingly, we might have good predictions because the closeness centrality would be different among nodes.

It is also seen that combination of global and local features did not necessarily lead to (considerable) improvement of the prediction accuracy. Though we have not yet identified the reason, this might be caused by overfitting. Identification of the reason is left as future work as well as introduction of some techniques to avoid overfitting.

As discussed, the results of computational experiments suggest that a combination of SVMs and structural features is useful for the prediction of nutritional requirements. Since nutritional requirements inferred by a scope-based method [8] are not necessarily perfect, it is worthy to develop alternative methods.

A combination of SVMs and structural features might be used as a complementary method to the scope-based method.

Though we have used artificially generated nutritional profiles, real nutritional requirements might be obtained by biological experiments. Therefore, use of real nutritional profiles is an important task for the future and requires a close collaboration with experimental biologists. Additional important future work is to improve the prediction accuracy. For that purpose, we need to develop novel structural features (especially local features). In particular, the introduction of local features defined for multiple nodes might be useful because there are many cases that knock-out of a single node (corresponding to a single enzyme/gene) does not affect functions of metabolic networks (because of the robustness of metabolic networks), but knock-out of multiple nodes greatly affects functions [5].

Acknowledgments

This work was partially supported by the International Research Training Group "Genomics and Systems Biology of Molecular Networks" Germany, and ITP (International Training Program) from JSPS, Japan.

References

[1] Albert, R., Barabási, A-L., Statistical mechanics of complex networks, *Rev. Mod. Phys.*, 74:47–97, 2002.

[2] Arita, M., The metabolic world of Escherichia coli is not small, *Proc. Natl. Acad. Sci. USA.*, 101:1543–1547, 2004.

[3] Barabási, A-L., Oltvai, Z. N., Network biology: understanding the cells' functional organization, *Nature Reviews Genetics*, 5:101–113, 2004.

[4] Bonacich, P., Some unique properties of eigenvector centrality, *Social Networks*, 29:555–564, 2007.

[5] Burgard, A. P., Pharkya, P., Maranas, C. D., OptKnock: A bilevel programming framework for identifying gene knockout strategies for microbial strain optimization, *Biotechnology and Bioenginerring*, 84:647–657, 2003.

[6] Cortes, C., Vapnik, V., Support-vector networks, *Machine Learning*, 20:273–297, 1995.

[7] Freeman, L. C., Centrality in social networks: Conceptual clarification, *Social Networks*, 1:215–239, 1979.

[8] Handorf, T., Christian, N., Ebenhöh, O., Kahn, D., An environmental perspective on metabolism, *Journal of Theoretical Biology*, 252:530–537, 2008.

[9] Handorf, T., Ebenhöh, O., Heinrich, R., Expanding metabolic networks: scopes of compounds, robustness, and evolution, *Journal of Molecular Evolution*, 61:498–512, 2005.

[10] Jeong, H., Tombor, B., Albert, R., Oltvai, Z. N., Barabási, A-L., Lethality and centrality in protein networks, *Nature*, 411:41–42, 2001.

[11] Kanehisa, M., Araki, M., Goto, S., Hattori, M., Hirakawa, M., Itoh, M., Katayama, T., Kawashima, S., Okuda, S., Tokimatsu, T., Yamanishi, Y., KEGG for linking genomes to life and the environment, *Nucleic Acids Research*, 36:D480–D484, 2008.

[12] Karp, P. D., Keseler, I. M., Shearer, A., Latendresse, M., Krummenacker, M., Paley, S. M., Paulsen, I., Collado-Vides, J., Gama-Castro, S., Peralta-Gil, M.,

Santos-Zavaleta, A., Penaloza-Spinola, M. I., Bonavides-Martinez, C., Ingraham, J., Multidimensional annotation of the Escherichia coli K-12 genome, *Nucleic Acids Research*, 35:7577–7590, 2007.

[13] Kim, H-J., Kim, J. M., Cyclic topology in complex networks, *Phys. Rev. E*, 72:036109, 2005.

[14] Matthews, B. W., Comparison of the predicted and observed secondary structure of T4 phage lysozyme, *Biochem. Biophys. Acta.*, 405:442–451, 1975.

[15] Newman, M. E. J., Assortative mixing in networks, *Phys. Rev. Lett.*, 89:208701, 2002.

[16] Newman, M. E. J., Power laws, Pareto distributions and Zipf's law, *Contemporary Physics*, 46:323–351, 2005.

[17] Pavlidis, P., Wapinski, I., Noble, W.S., Support vector machine classification on the web, *Bioinformatics*, 20(4):586–587, 2004.

[18] Papin, J. A., Stelling, J., Price, N. D., Klamt, S., Schuster, S., Palsson, B. O., Comparison of network-based pathway analysis methods, *TRENDS in Biotechnology*, 22:400–405, 2004.

[19] Shawe-Taylor, J., Cristianini, N., Kernel Methods for Pattern Analysis, *Cambridge Univ. Press*, 2004.

[20] Stelling, J., Klamt, S., Bettenbrock, K, Schuster, S., Gilles, E. D., Metabolic network structure determines key aspects of functionality and regulation, *Nature* 420:190–193, 2002.

[21] Takemoto, K., Nacher, J. C., Akutsu, T., Correlation between structure and temperature in prokaryotic metabolic networks, *BMC Bioinformatics*, 8:303, 2007.

[22] Vázquez, A., Dobrin, R., Sergi, D., Eckmann, J. P., Oltvai, Z. N., Barabási, A. L,, The topological relationship between the large-scale attributes and local interaction patterns of complex networks, *Proc. Natl. Acad. Sci. USA*, 101:17940–17945, 2004.

[23] Wagner, A., Fell, D., The small world inside large metabolic networks, *Proceedings of the Royal Society of London B*, 268:1803–1810, 2001.

ANALYSIS OF A LIPID BIOSYNTHESIS PROTEIN FAMILY AND PHOSPHOLIPID STRUCTURAL VARIATIONS

MICHIHIRO TANAKA[1]
mtanaka@kuicr.kyoto-u.ac.jp

YUKI MORIYA[1]
moriya@kuicr.kyoto-u.ac.jp

SUSUMU GOTO[1]
goto@kuicr.kyoto-u.ac.jp

MINORU KANEHISA[1,2]
kanehisa@kuicr.kyoto-u.ac.jp

[1] *Bioinformatics Center, Institute for Chemical Research, Kyoto University, Uji, Kyoto 611-0011, Japan*

[2] *Human Genome Center, Institute of Medical Science, University of Tokyo, Minato-ku, Tokyo 108-8639, Japan*

Glycerophospholipids are major structural lipids in cellular membrane systems and play key roles as suppliers of the first and second messengers in the signal transduction and molecular recognition processes. The distribution of lipid components differs among organelles and cells. The distribution is controlled by two pathways in lipid metabolism: *de novo* and remodeling pathways. Glycerophospholipids including arachidonic and stearic acids are mostly produced in the remodeling pathway, whereas lipid chains are reconstructed from those synthesized in the *de novo* pathway. Recently lysophospholipid acyltransferases have been isolated as key enzymes in the remodeling pathway, and the substrate specificity has been investigated in terms of the chemical substructures of glycerophospholipids, such as the type of head groups and the length of aliphatic chains. These experimental studies have been reported for specific organisms, and only two representative sequence motifs are known for acyltransferases: a general pattern and the pattern for membrane-bound *O*-acyltransferase (MBOAT). Here we attempt to correlate the sequence patterns and the substrate specificity of lysophospholipid acyltransferases in 89 eukaryotic genomes in order to understand the roles of this enzyme family and underlying glycerophospholipid structural variations. Using phylogenetic and domain analyses, the lysophospholipid acyltransferase family was divided into 18 subtypes. Furthermore, we examined the occurrence of identified subtypes in eukaryotic genomes, and found the expansion of these subtypes in vertebrates. These findings may provide clues to understanding structural variations and distributions of glycerophospholipids in different organisms.

Keywords: lysoacyltransferase; glycerophospholipids remodeling pathways.

1. Introduction

Glycerophospholipids are major structural lipids in cellular membrane systems and play key roles as suppliers of the first and second messengers in the signal transduction and molecular recognition processes [1]. They have several molecular variations stemming from the combination of three substructure components: a polar-head group and two aliphatic chains at *sn*-1 and *sn*-2 positions. Based on the polar-head group, glycerophospholipids are classified into several classes, such as phosphatidic acid (PA), phosphatidylcholine (PC), phosphatidylethanolamine (PE), phosphatidylglycerol (PG), cardiolipin (CL), phosphatidylinositol (PI) and phosphatidylserine (PS). Combined with the variation of two aliphatic chains, over 1,000 glycerophospholipids species have been reported so far [1].

The distribution of lipid species differs among organelles and cells. Glycerophospholipid biosynthesis is controlled by two lipid metabolism pathways: the *de novo* pathway [2] and the remodeling pathway [3]. Two acyltransferases, glycerol-phosphate acyltransferase (GPAT) and acylglycerol-phosphate acyltransferase (AGPAT), are responsible for acylation in the *de novo* pathway. Because these enzymes have no substrate specificities to aliphatic chains [4-7], the biased distributions of lipid species are not generated in the *de novo* pathway. The remodeling pathway uses phospholiase A_2 (PLA_2) and lysophospholipid acyltransferases (LPLATs). PLA_2 hydrolyses the ester bond at *sn*-2 position of a glycerophospholipid. The produced free fatty acid and lysoacylglycerophospholipids are then used as precursors for various second messengers. LPLAT converts the remaining lysoacylgycerophospholipid into another glycerophospholipid, in which the aliphatic chain at *sn*-2 position may differ from that of the original one. For example, glycerophospholipids including arachidonic and stearic acids are mostly produced in the remodeling pathway [8]. Many members of the PLA_2 gene family have been cloned, and their substrate specificities for aliphatic chains have been well characterized, as well as the corresponding sequence motifs [9-12]. On the other hand, only several members from the mammalian LPLAT family have been isolated [13-16, 17-20], although their substrate specificity has been investigated in terms of the chemical substructures of glycerophospholipids, such as the type of head groups and the length of aliphatic chains [20]. However, these experimental studies have been performed only for certain organisms, and only two representative sequence motifs are currently registered in the Pfam motif database [21]: a general acyltransferase pattern (Acyltransferase) and the pattern for membrane-bound *O*-acyltransferase (MBOAT).

Here we attempt to correlate the sequence patterns and the substrate specificity of LPLATs in 89 eukaryotic genomes in order to understand the roles of this enzyme family and underlying glycerophospholipid structural variations. Using phylogenetic and domain analyses, the LPLAT family was divided into 18 subtypes. Furthermore, we examined the occurrence of identified subtypes in the eukaryotic genomes and found that these subtypes are expanded in vertebrates. These findings may provide clues to understanding structural variations and distributions of glycerophospholipids in different organisms.

2.   Materials and Methods

2.1.   *Experimentally characterized lysophospholipid acyltransferase sequences*

We collected all possible LPLATs (lysophospholipid acyltransferases) from completely sequenced eukaryotes to analyze species variations of glycerophospholipids (Fig. 1). To provide initial query sequences, we retrieved experimentally characterized LPLAT encoding protein sequences from the UniProtKB database [22] using several keywords related to lysophospholipid acyltransferase followed by manual inspection for substrate specificities. We obtained 24 sequences, including 7 sequences from *H. sapiens* (Q6P1A2, A9EDR2, A9EDQ5, Q6UWP7, Q7L5N7, Q96N66 and Q8NF37), 6 sequences from *M. musculus* (Q91V01, Q3TFD2, Q8BH98, Q3UN02, Q8BYI6 and Q8R3I2), 3 sequences from *D. rerio* (Q6NYV8, Q502J0 and Q1LWG4), 2 sequences from *R. norvegicus* (Q1HAQ0 and Q5FVN0), and a sequence from each of *B. taurus* (Q3SZL3), *G. gallus* (Q5F3X0), *D. melanogaster* (Q0KHU5), *S. cerevisiae* (Q06510), *S. pombe* (O42916) and *D. discoideum* (Q54DX7).

2.2. *PSI-BLAST Search for lysophospholipid acyltransferase candidates*

We performed PSI-BLAST search [23] on eukaryotes in the KEGG GENES database [24] using each of the above 24 sequences as a seed. The following two conditions were used for each PSI-BLAST search:

i) The E-value threshold was set to $< 10^{-2}$ for creating the sequence set to generate a position-specific scoring matrix (PSSM).

ii) The PSI-BLAST hit should include the seed sequence, and the maximum number of iteration cycles was set to 30.

iii) All of the results were integrated to produce a list of non-redundant sequences.

2.3. *Clustering analysis*

We divided PSI-BLAST hits into several clusters to identify non-LPLAT sequences (false positives) in the following way:

i) A Smith-Waterman score matrix was constructed using all-against-all SSEARCH [17] for the sequences in the list.

ii) The complete linkage method was applied to the score matrix by using the R package [26] with the threshold Smith-Waterman score > 20.

LPLATs are known to have either of two Pfam motifs, PF01553 (Acyltransferase) or PF03062 (MBOAT). We considered the clusters that did not contain either of these motifs as false positives and did not use them for further analysis.

2.4. *Phylogenetic analysis of each cluster*

To determine the orthologous relationships among sequences, we applied the tree-reconciliation phylogenetic approach [27] to each of the remaining clusters.

i) We used *E. coli* acyltransferase (eco:b2836) as the outgroup of lysophospholipid acyltransferase family to construct rooted gene trees.

ii) For each cluster, a multiple sequence alignment was created using MAFFT [28] with default parameters except for max-iteration, which was set to 30.

iii) If more than 10% of the sequences in an alignment region corresponded to gaps, the region was eliminated.

iv) Neighbor-joining phylogenies were constructed using the NEIGHBOR program in the PHYLIP 3.6 package [29]. One hundred bootstrap replicates were generated to assess the variation in the data using the PHYLIP 3.6 SEQBOOT program.

v) A species tree for the 89 eukaryotes was constructed based on KEGG taxonomy data [30] and a previously defined metazoan species tree [31]. The list of 89 eukaryotic species and taxonomic classification used in this study can be obtained from http://mbi3.kuicr.kyoto-u.ac.jp/supp/mtanaka/taxonomy_ibsb2009.txt

To detect duplication and speciation events on the gene trees, we used a phylogeny-based algorithm [27]. This algorithm fits a given gene tree to its corresponding species tree and then infers the minimum set of duplications necessary to explain the topology of the gene tree. We call the derived tree with the duplication events a reconciled tree. We manually investigated the reconciled tree to infer the orthologous and paralogous relationships for sequence pairs in a given cluster.

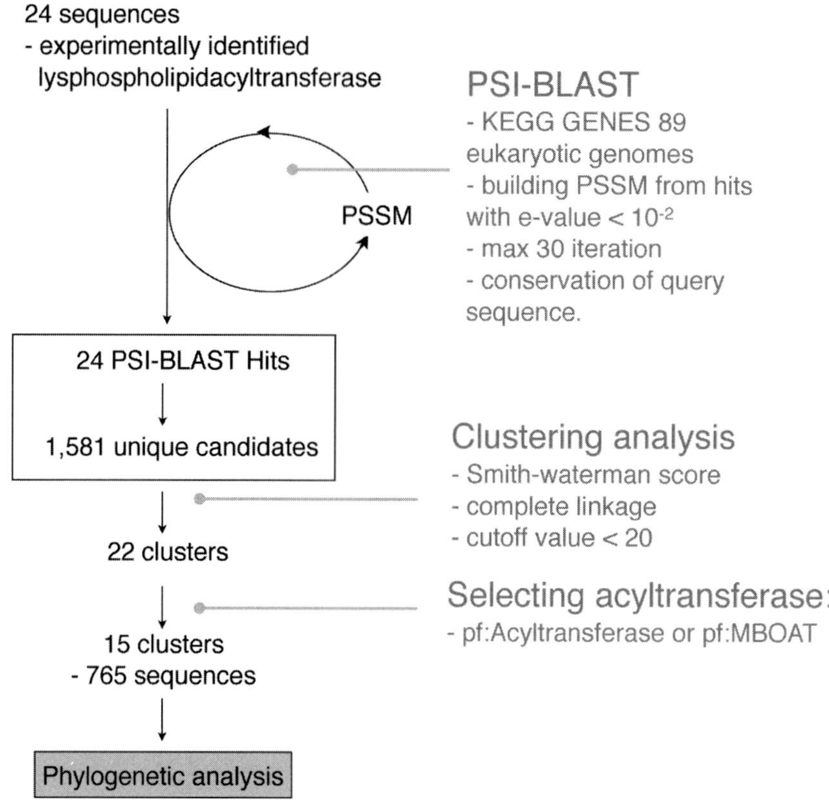

Figure 1. The protocol for detecting lysophospholipid acyltransferase candidates.

3.   Results

3.1.   *Identification of lysophospholipid acyltransferase sequences*

We obtained 1,581 LPLAT candidates among 89 eukaryotes through PSI-BLAST. The successive clustering analysis divided them into 22 clusters. After examining the occurrence of two Pfam motifs (pf:Acyltransferase and pf:MBOAT) for each of 22 clusters, 7 clusters were considered as false positives and excluded from further analyses. Of the remaining 15 clusters, 9 clusters (429 sequences from 75 species) contained sequences with the Acyltransferase motif and the other 6 clusters (336 sequences from 75 species) contained sequences with the MBOAT motif. In total, we obtained 765 LPLAT sequences in 89 eukaryotic genomes in the 15 clusters.

3.2. *Phylogenetic analysis for LPLAT sequences*

The taxonomic distribution of the LPLATs in the 15 clusters is shown in Table 1. Note that genes belonging to a cluster are sometimes highly duplicated in a specific taxonomy. For example, Deuterostomia has 3.94 sequences per species in cluster A, which is the largest cluster within the Acyltransferase family, and other taxonomies (from Protostomia to Plants) have 2.06 to 3.33 sequences per species in the cluster. Another example is cluster J, which is the largest cluster within the MBOAT family. Deuterostomia has 2.56 sequences per species, but other taxonomies have 1.00 to 1.88 sequences per species in this cluster. These results suggest that the Metazoa-specific and Deuterostomia-specific gene expansions have occurred within the Acyltransferase family and MBOAT family, respectively.

Table 1. Taxonomic distribution of the LPLAT gene families

Cluster	# of sequences	# of subtypes	Deuterostomia(18)		Protostomia(19)		Cnidarians(1)		Placozoans(1)		Plants(6)		Fungi(26)		Protists(18)	
Acyltransferase																
-A	144	3	**3.94**	(71/18)	**2.06**	(37/18)	**3.00**	(3/1)	**3.00**	(3/1)	**3.33**	(10/3)	1.15	(15/13)	1.67	(5/3)
-B	92	3	**2.00**	(34/17)	1.63	(31/19)	1.00	(1/1)	**2.00**	(2/1)	**3.33**	(20/6)	-		0.75	(3/4)
-C	67	4	**2.76**	(47/17)	1.00	(16/16)	**2.00**	(2/1)	**2.00**	(2/1)	-		-		-	
-D	53	1	1.13	(18/16)	1.00	(19/19)	1.00	(1/1)	1.00	(1/1)	1.83	(11/6)	1.00	(2/2)	1.00	(1/1)
-E	23	1	-		-		-		-		-		1.05	(22/21)	1.00	(1/1)
-F	22	1	1.00	(7/7)	1.00	(12/12)	-		-		-		1.00	(3/3)	-	
-G	15	1	1.08	(14/13)	1.00	(1/1)	-		-		-		-		-	
-H	7	1	-		-		-		-		-		1.00	(7/7)	-	
-I	6	1	**4.00**	(4/1)	1.00	(3/3)	-		-		-		1.00	(2/2)	-	
MBOAT																
-J	98	2	**2.56**	(41/16)	1.88	(30/16)	1.00	(1/1)	-		1.00	(1/1)	1.60	(24/15)	1.00	(1/1)
-K	80	5	**2.77**	(61/22)	1.00	(19/19)	1.00	(1/1)	1.00	(1/1)	-		-		**3.50**	(7/2)
-L	69	1	1.00	(3/3)	1.16	(22/19)	**2.00**	(2/1)	-		1.60	(8/5)	1.08	(27/25)	**2.33**	(7/3)
-M	43	1	1.31	(21/16)	1.18	(20/17)	-		**2.00**	(2/1)	-		-		-	
-N	31	1	1.00	(8/8)	-		-		-		-		1.10	(23/21)	-	
-O	15	1	1.21	(17/14)	1.00	(1/1)	-		-		-		-		-	

Bold numbers show taxonomies with ≥ 2 sequences per species in the cluster.
Two numbers in the parenthesis show the number of sequences and species, respectively.

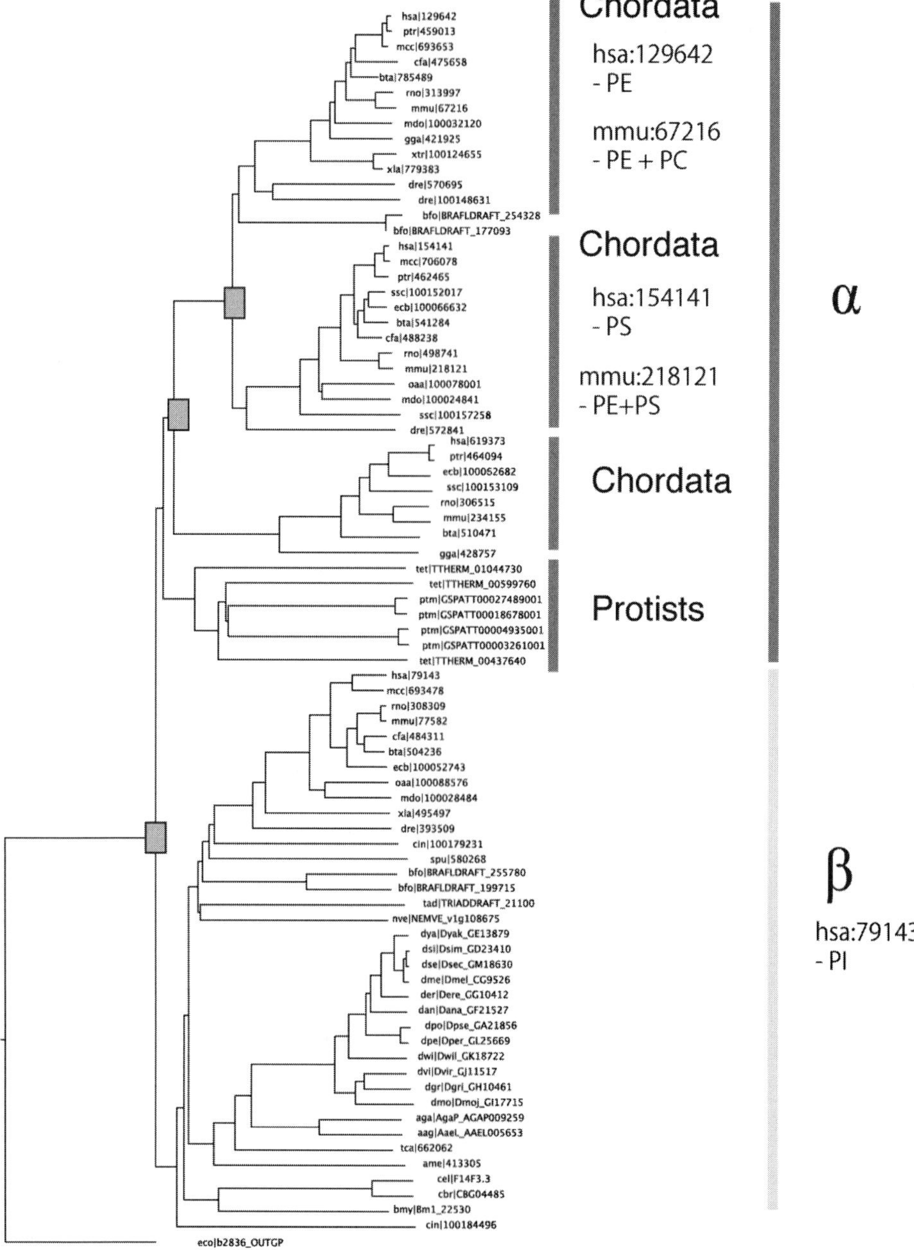

Figure 2. Phylogenetic tree representing the evolutionary relationships among sequences in cluster K. The branches representing duplication events are marked as grey squares.

In contrast, several clusters show a tandem distribution for a number of genes among different species. Deuterostomia, Protostomia and Fungi each have 1.00 sequence per species in clusters F and G, while other taxonomies have no sequences in these clusters. Similar distributions can be observed in clusters N and M. They also show sparse distributions along the wide range of taxonomies, suggesting that gene loss events have occurred frequently in these clusters.

To clarify the evolutionary process within the 15 clusters, we further examined and classified orthologous and paralogous relationships of sequence pairs by manually investigating the reconciled trees that derived from the gene trees. The gene trees for the 15 clusters were supported by bootstrap values > 45%. The number of subtypes in each cluster by this classification is shown in Table 1. Figure 2 shows the phylogenetic relationships of cluster K. With this result, we estimated that cluster K was divided into two subtypes (α and β), and one of the subtypes was further divided into three Chordata subtypes and a Protist subtype. Therefore, this reconciled tree indicates that several gene duplication events have occurred in the Chordata lineage, and each subtype may have evolved to have the substrate specificity of LPLATs.

3.3. *Applications of orthologous relationships to inferring the function of LPLATs*

The branching of gene trees emerges from two events. One is the speciation event. If a branching point between genes in a given gene tree derived from a speciation event, we consider this as an orthologous relationship. The function is usually conserved between orthologous genes. The other type of event is the ancestral gene duplication event that leads to a paralogous relationship between the duplicated genes. Paralogous genes can evolve different functions. To infer the function of a given gene from the other functionally characterized genes using phylogeny-based approach, we should correctly distinguish between orthologous and paralogous relationships for each branching point in a given tree.

To infer functions of LPLATs of genes with unknown function or substrate specificity, we used orthologous relationships of the 24 experimentally characterized LPLATs that were in 7 clusters among the 15 clusters. The 7 clusters included 4 clusters from Acyltransferase families (clusters A, C, D, E) and 3 clusters from MBOAT families (clusters K, L, M), and 479 LPLAT sequences in total. Table 2 shows the phylogenetic distributions of the LPLAT subtypes for which polar-head group specificity can be inferred from the experimentally characterized LPLATs.

4. Discussion

In this work, we classified lysophospholipid acyltransferase sequences hierarchically and also examined the relationship between subtypes and polar-head group specificity. However, we could not identify LPLAT subtypes for PS, PG, PI and CL in plants and fungi. This may be due to 24 biased query sequences that include only two fungi and no plant sequences. Therefore, we need to find such LPLAT sequences. Because we did not use several clusters in Table 1 for the substrate specificity analysis, these sequences may be classified in the uninvestigated clusters. Further investigation of the sequences in the subtypes in Table 2 may reveal novel features for the substrate specificities, which can be applied to the prediction of the function of LPLAT in those clusters.

It is difficult to elucidate the variation of glycerophospholipids only by the results for LPLAT analysis in this study. For example, Substrate specificifies of LPLAT have been reported in terms of the length of aliphatic chains [20]. However, we have not investigated them in the current study, and to examine them in detail is one of our future works.

To better understand variation of the synthesis mechanisms of glycerophospholipids, we need to consider at least four other enzymes: elongases [32], desaturases [33], acyl-CoA synthetases [34] and phospholipases [35]. Desaturases and elongases play important roles in *de novo* fatty acid synthesis. Hashimoto *et al* reported that fatty acid variations could be predicted from the combination of subfamilies for desaturases and elongases [36, 37]. An important step in transferring fatty acids to the glycerol backbone is acyl-CoA synthesis. Variation in the acyl-CoA synthetase family has been reported based on the substrate specificity of the chain length of fatty acids [9-11]. Phospholipases are also classified into over 20 subfamilies whose acyl chain selectivities are already known. For example, cPLA2 α has selectivity for arachidonic acid, and iPLA2 β has no selectivity for specific acyl chains (a general purpose lipase). We must comprehensively understand the relationship between these enzymes and genetic variation among their encoding genes to elucidate the variation of glycerophospholipids in cellular membranes.

Table 2. Phylogenetic distribution of LPLAT subtypes based on head-group specificity of the substrate glycerophospholipids. Subtypes consist of a cluster ID and a subtype number. For example, C1 represents the first subtype of the cluster C. PC: Phosphatidylcholine, PE: Phosphatidylethanolamine, PS: Phosphatidylserine, PG: Phosphatidylglycerol, PI: Phosphatidylinositol, CL: Cardiolipin

Head group	Class	Vertebrates	Protochordata	Echinodemata	Protostomia	Cnidarians	Placozoans	Plants	Fungi	Protists
PC										
	- C1	+	+	+	+	+				
	- C2	+								
	- D	+	+	+	+	+	+	+	+	+
	- E								+	+
	- K1	+	+							
	- L	+	+	+	+	+		+	+	+
	- M	+	+	+	+		+			
PE										
	- K1	+	+							
	- K2	+								
	- K3	+								
	- K4									+
	- L	+	+	+	+	+		+	+	+
PS										
	- K2	+								
	- M	+	+	+	+		+			
PG	A3	+	+	+	+	+	+			+
PI										
	- A3	+	+	+	+	+	+			+
	- K5	+	+	+	+	+	+			
CL	A3	+	+	+	+	+	+			+

In conclusion, we have classified 765 lysophospholipid acyltransfease candidates into 15 clusters and defined 18 lysophospholipid acyltransferase subtypes for 7 of 15 clusters through phylogenetic analysis. Further analyses of these results may provide new insights into mechanisms contributing to the variation of glycerophospholipids.

Acknowledgments

We thank Dr. Kosuke Hashimoto and Dr. Nelson Hayes for critical reading of our manuscript. This work was supported in part by the grants from the Ministry of Education, Culture, Sports, Science and Technology. The computational resource was provided by the Bioinformatics Center, Institute for Chemical Research, Kyoto University.

References

[1]   van Meer, G., Voelker, D.R., Feigenson, G.W., Membrane lipids: where they are and how they behave. *Nat Rev Mol Cell Biol*, 9(2):112-124, 2008.

[2]   Kennedy, E.P., Weiss, S.B., The function of cytidine coenzymes in the biosynthesis of phospholipides. *J Biol Chem*, 222(1):193-214, 1956.

[3]   Lands, W.E., Metabolism of glycerolipides; a comparison of lecithin and triglyceride synthesis. *J Biol Chem*, 231(2):883-888, 1958.

[4]   Holub, B.J., Kuksis, A., Metabolism of molecular species of diacylglycerophospholipids. *Adv Lipid Res*, 16:1-125, 1978.

[5]   Miki, Y., Hosaka, K., Yamashita, S., Handa, H., Numa, S., Acyl-acceptor specificities of 1-acylglycerolphosphate acyltransferase and 1-acylglycerophosphorylcholine acyltransferase resolved from rat liver microsomes. *Eur J Biochem*, 81(3):433-441, 1977.

[6]   Waku, K., Origins and fates of fatty acyl-CoA esters. *Biochim Biophys Acta*, 1124(2):101-111, 1992.

[7]   Nakagawa, Y., Rustow, B., Rabe, H., Kunze, D., Waku, K., The de novo synthesis of molecular species of phosphatidylinositol from endogenously labeled CDP diacylglycerol in alveolar macrophage microsomes. *Arch Biochem Biophys*, 268(2):559-566, 1989.

[8]   Sugiura, T., Masuzawa, Y., Nakagawa, Y., Waku, K., Transacylation of lyso platelet-activating factor and other lysophospholipids by macrophage microsomes. Distinct donor and acceptor selectivities. *J Biol Chem*, 262(3):1199-1205, 1987.

[9]   Soupene, E., Kuypers, F.A., Mammalian long-chain acyl-CoA synthetases. *Exp Biol Med* (Maywood), 233(5):507-521, 2008.

[10] Fujino, T., Takei, Y.A., Sone, H., Ioka, R.X., Kamataki, A., Magoori, K., Takahashi, S., Sakai, J., Yamamoto, T.T., Molecular identification and characterization of two medium-chain acyl-CoA synthetases, MACS1 and the Sa gene product. *J Biol Chem*, 276(38):35961-35966, 2001.

[11] Watkins, P.A., Very-long-chain acyl-CoA synthetases. *J Biol Chem*, 283(4):1773-1777, 2008.

[12] Kratzer, S., Schuller, H.J., Carbon source-dependent regulation of the acetyl-coenzyme A synthetase-encoding gene ACS1 from Saccharomyces cerevisiae. *Gene*, 161(1):75-79, 1995.

[13] Kume, K., Shimizu, T., cDNA cloning and expression of murine 1-acyl-sn-glycerol-3-phosphate acyltransferase. *Biochem Biophys Res Commun*, 237(3):663-666, 1997.

[14] Eberhardt, C., Gray, P.W., Tjoelker, L.W., Human lysophosphatidic acid acyltransferase. cDNA cloning, expression, and localization to chromosome 9q34.3. *J Biol Chem*, 272(32):20299-20305, 1997.

[15] Hollenback, D., Bonham, L., Law, L., Rossnagle, E., Romero, L., Carew, H., Tompkins, C.K., Leung, D.W., Singer, J.W., White, T., Substrate specificity of lysophosphatidic acid acyltransferase beta - evidence from membrane and whole cell assays. *J Lipid Res*, 47(3):593-604, 2006.

[16] Yang, Y., Cao, J., Shi, Y., Identification and characterization of a gene encoding human LPGAT1, an endoplasmic reticulum-associated lysophosphatidylglycerol acyltransferase. *J Biol Chem*, 279(53):55866-55874, 2004.

[17] Nakanishi, H., Shindou, H., Hishikawa, D., Harayama, T., Ogasawara, R., Suwabe, A., Taguchi, R., Shimizu, T., Cloning and characterization of mouse lung-type acyl-CoA:lysophosphatidylcholine acyltransferase 1 (LPCAT1). Expression in alveolar type II cells and possible involvement in surfactant production. *J Biol Chem*, 281(29):20140-20147, 2006.

[18] Shindou, H., Hishikawa, D., Nakanishi, H., Harayama, T., Ishii, S., Taguchi, R., Shimizu, T., A single enzyme catalyzes both platelet-activating factor production and membrane biogenesis of inflammatory cells. Cloning and characterization of acetyl-CoA:LYSO-PAF acetyltransferase. *J Biol Chem*, 282(9):6532-6539, 2007.

[19] Chen, X., Hyatt, B.A., Mucenski, M.L., Mason, R.J., Shannon, J.M., Identification and characterization of a lysophosphatidylcholine acyltransferase in alveolar type II cells. *Proc Natl Acad Sci U S A*, 103(31):11724-11729, 2006.

[20] Hishikawa, D., Shindou, H., Kobayashi, S., Nakanishi, H., Taguchi, R., Shimizu, T., Discovery of a lysophospholipid acyltransferase family essential for membrane asymmetry and diversity. *Proc Natl Acad Sci U S A*, 105(8):2830-2835, 2008.

[21] Finn, R.D., Tate, J., Mistry, J., Coggill, P.C., Sammut, S.J., Hotz, H.R., Ceric, G., Forslund, K., Eddy, S.R., Sonnhammer, E.L., et al., The Pfam protein families database. *Nucleic Acids Res*, 36(Database issue):D281-288, 2008.

[22] http://www.uniprot.org/help/uniprotkb.

[23] Altschul, S.F., Madden, T.L., Schaffer, A.A., Zhang, J., Zhang, Z., Miller, W., Lipman, D.J., Gapped BLAST and PSI-BLAST: a new generation of protein database search programs. *Nucleic Acids Res*, 25(17):3389-3402, 1997.

[24] Kanehisa, M., Araki, M., Goto, S., Hattori, M., Hirakawa, M., Itoh, M., Katayama, T., Kawashima, S., Okuda, S., Tokimatsu, T., et al., KEGG for linking genomes to life and the environment. *Nucleic Acids Res*, 36(Database issue):D480-484, 2008.

[25] Smith, T.F., Waterman, M.S., Identification of common molecular subsequences. *J Mol Biol*, 147(1):195-197, 1981.

[26] Team RDC: R: a language and environment for statistical computing Vienna, Austria: R Foundation for Statistical Computing 2003.

[27] Goodman, M.C.J., Moore, G.M., Romero-Herrera, A.E., Matsuda, G., Fitting the gene lineage into its species lineage, a parsimony strategy illustrated by cladograms constructed from globin sequences. *Syst Zool*, 28:132-163, 1979.

[28] Katoh, K., Kuma, K., Toh, H., Miyata, T., MAFFT version 5: improvement in accuracy of multiple sequence alignment. *Nucleic Acids Res*, 33(2):511-518, 2005.

[29] Felsenstein, J., PHYLIP: Phylogenetic Inference Package. Release 3.6. Department of Genome Sciences, University of Washington, Seattle. 2005.

[30] ftp://ftp.genome.jp/pub/kegg/genes/taxonomy.

[31] Ponting, C.P., The functional repertoires of metazoan genomes. *Nat Rev Genet*, 9(9):689-698, 2008.

[32] Leonard, A.E., Pereira, S.L., Sprecher, H., Huang, Y.S., Elongation of long-chain fatty acids. *Prog Lipid Res*, 43(1):36-54, 2004.

[33] Sperling, P., Ternes, P., Zank, T.K., Heinz, E., The evolution of desaturases. *Prostaglandins Leukot Essent Fatty Acids*, 68(2):73-95, 2003.

[34] Watkins, P.A., Fatty acid activation. *Prog Lipid Res*, 36(1):55-83, 1997.

[35] Burke, J.E., Dennis, E.A., Phospholipase A2 structure/function, mechanism, and signaling. J Lipid Res, 50 Suppl:S237-242, 2009.

[36] Hashimoto, K., Yoshizawa, A.C., Okuda, S., Kuma, K., Goto, S., Kanehisa, M., The repertoire of desaturases and elongases reveals fatty acid variations in 56 eukaryotic genomes. *J Lipid Res*, 49(1):183-191, 2008.

[37] Hashimoto, K., Yoshizawa, A.C., Saito, K., Yamada, T., Kanehisa, M., The repertoire of desaturases for unsaturated fatty acid synthesis in 397 genomes. *Genome Inform*, 17(1):173-183, 2006.

CaMPDB: A RESOURCE FOR CALPAIN AND MODULATORY PROTEOLYSIS

DAVID duVERLE[1]   ICHIGAKU TAKIGAWA[1]
dave@kuicr.kyoto-u.ac.jp   takigawa@kuicr.kyoto-u.ac.jp

YASUKO ONO[2]   HIROYUKI SORIMACHI[2]   HIROSHI MAMITSUKA[1]
ono-ys@igakuken.or.jp   sorimachi-hr@igakuken.or.jp   mami@kuicr.kyoto-u.ac.jp

[1] *Bioinformatics Center, Kyoto University, Gokasho, Uji 611-0011, Japan*
[2] *Calpain Project, Rinshoken, Tokyo 156-8506, Japan*

While the importance of modulatory proteolysis in research has steadily increased, knowledge on this process has remained largely disorganized, with the nature and role of entities composing modulatory proteolysis still uncertain. We built CaMPDB, a resource on modulatory proteolysis, with a focus on calpain, a well-studied intracellular protease which regulates substrate functions by proteolytic processing. CaMPDB contains sequences of calpains, substrates and inhibitors as well as substrate cleavage sites, collected from the literature. Some cleavage efficiencies were evaluated by biochemical experiments and a cleavage site prediction tool is provided to assist biologists in understanding calpain-mediated cellular processes. CaMPDB is freely accessible at
http://calpain.org

Keywords: calpain; calpastatin; database; modulatory proteolysis.

1. Introduction

Proteolysis is recognized as the fourth step of the central dogma, next to the basic three stages: replication, transcription and translation. Proteolysis generally indicates a process of protein degradation by a proteolytic enzyme. Recently, it has been shown that a certain type of proteolysis, *modulatory proteolysis*, regulates substrate functions by substrate-specific (or amino acid position-specific) proteolysis. It further suggests that modulatory proteolysis controls almost all types of cellular functions [3, 7, 16].

While there are many databases on proteases [6, 12] which give a broad view of protease sequences, CaMPDB is built as a resource for modulatory proteolysis, focusing on calpain, a Ca^{2+}-dependent Cys protease. Calpains constitute a major protease family distributed over a wide range of organisms, from mammals to bacteria, for which many isoforms and homologues are being found [2, 15]. More importantly, calpains regulate substrate functions by limited proteolysis, *i.e.* proteolytic processing, resulting in the modulation of a wide variety of biological phenomena. Malfunction of calpain has been observed in several serious diseases, including muscular dystrophies and diabetes [1]. Activity of calpain in vivo is regulated by its specific endogenous inhibitor protein, calpastatin [3]. Although calpastatin binds to calpain around the catalytic site cleft like regular substrates, it is resistant to proteolysis by calpain (see Fig. 6 for details).

The molecular entities composing modulatory proteolysis are still largely unknown, mainly due to disorganized knowledge on this process. Generating a repository of information on calpain-mediated proteolysis therefore offers a number of advantages. In particular, precise and specific recognition of the cleavage site is critical in understanding modulatory proteolytic processing. Amino acid sequence specificity of cleavage by calpains, if it exists, has not been established to this day, although some preferences regarding amino acid residues close to the cleavage site were reported [9, 13, 17]. This makes it hard to predict whether a given protein can be proteolyzed by calpains and further hampers elucidation of physiological functions of calpains.

In light of this, we have built CaMPDB with the following unique features:

(1) Large collection of calpain sequences gathered from a variety of organisms, along with information on substrates (cleavage sites) and calpastatin, both of which interact with the catalytic center of calpains.
(2) Confirmation of collected cleavage sites through biochemical experiments: for a total of 56 cases, we were also able to present cleavage efficiency.
(3) Statistics based on collected data, such as amino acid preferences of cleavage sites or philogenetic analysis of calpastatin's functional domains across species.
(4) Built-in cleavage prediction tool that can take any arbitrary input sequence and score each position for putative cleavage by calpain. We believe such a feature will be of great help for understanding various calpain-mediated cellular processes.

2. Database Framework

CaMPDB has a total of 3,723 entries over 363 organisms, separated into three categories: 'Calpain', 'Substrates' and 'Calpastatin'. Each category is broken down into four standard subsections: 'Overview', 'Browse', 'Search' and 'Statistics', with an additional 'Predict' subsection for 'Substrates'. 'Overview' contains a detailed description on the subject of each category. For example, 'Overview' of 'Calpain' gives a detailed review of history, nomenclature, structure, functions, superfamily, references and glossary of calpain. 'Browse' provides a listing of entries stored for each category. Fig. 1 shows some screen-captures of entry SB0019: Interleukin 1α, a substrate of calpain.

Each entry includes basic nomenclature data (Fig. 1 (a)), cross-references to six major databases (Entrez, Swiss-Prot, OMIM, HGNC, HPRD and KEGG), information from OMIM, structure information consisting of: domain information (Fig. 1 (b)), amino acid sequence with computationally predicted secondary structures (Fig. 1 (c)), 3D structure and references linking to PubMed.

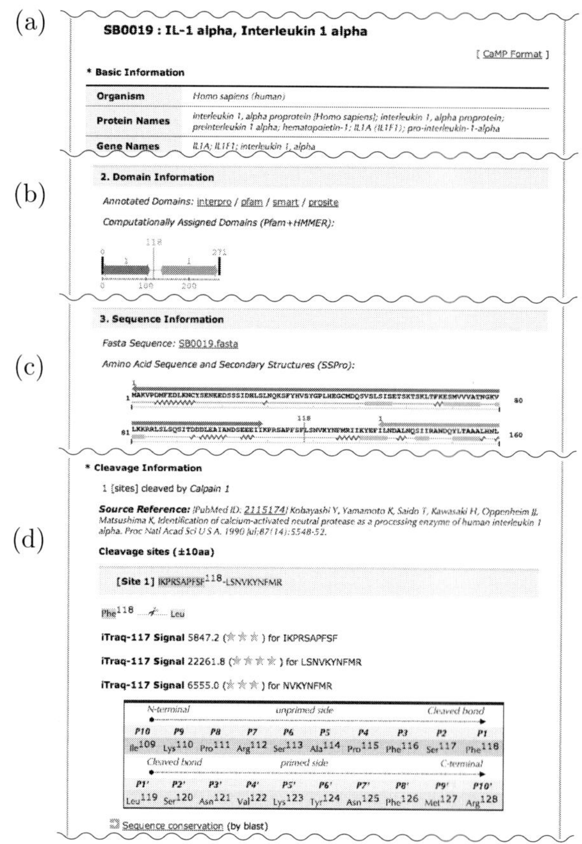

Fig. 1: Screen-captures of entry SB0019: (a) Basic, (b) domain, (c) sequence and (d) cleavage information.

It can be noted that each entry in 'Substrates' has some additional fields regarding cleavage sites (see section 3). Entries in each category can be retrieved in two ways: through browsing (as a flat list or a clickable taxonomy tree) or by query search.

In the 'Search' subsection, pertinent results for a specific query can be obtained for each category, while a separate form linked from the main page, enables searches over the entire database. The 'Search' interface allows filtering by name, query ID or any other text annotation.

'Statistics' shows a summary of statistics for the corresponding category along with some data analysis results: for the 'Calpain' category, we present a breakdown of 16 calpain types across 136 organisms. 'Substrates' shows position-specific amino acid preferences for cleavage sites using PSSM (Position-Specific Scoring Matrix, see Section 3) [5] and sequence logo [14], generated from 255 collected cleavage site sequences.

The subsection for 'Calpastatin' presents the evolution of its four known functional domains across 29 representative organisms as a phylogenetic tree (see Fig. 2).

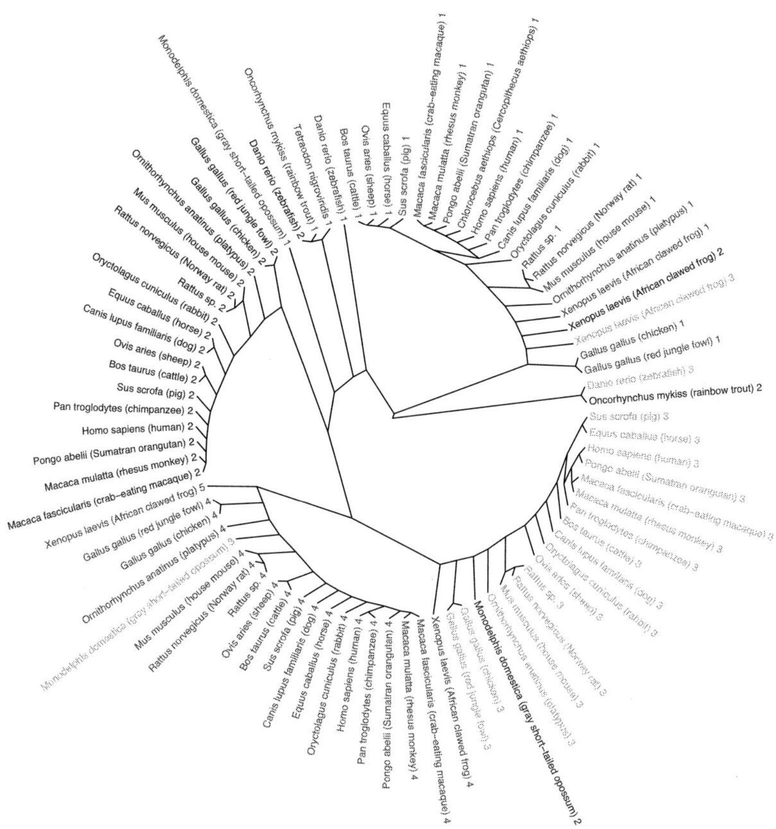

Fig. 2: Radial philogenetic tree of calpastatin functional domain sequences.

In addition, CaMPDB has separate pages for usage instructions, sitemap and downloading of all data.

2.1. *Data Collection*

For 'Calpain', we collected 1,496 entries from Entrez that were marked with *CysPc*, indicating Ca^{2+}-dependent cysteine proteinase superfamily.

For 'Substrates', we collected 104 known calpain substrates (SB) from literature as entries. Some SB had more than one cleavage sites, resulting in 267 cleavage sites being stored in CaMPDB. Each entry in 'Substrates' also contains, for each of cleavage site (Fig. 1 (d)): the cleavage site sequence and similar sequences across organisms that can be obtained by casting the cleavage site (with a length of 30) as a query to BLAST. The latter is very helpful for biologists to check how commonly the cleavage site can be found in other organisms and to infer biological phenomena with which calpains could be related.

Table 1: Summary of sequence data contained in the database.

Organism	CL	XB+XSB	SB	XSB	CT	**Total**
Bacteria	38	0	0	0	4	42
Eukaryota	1402	1988	104	1884	193	3583
Mammalia	571	1292	97	1195	178	2041
Others	831	696	7	689	15	1542
Unknown	56	30	0	30	12	98
Total	1496	2218	104	1914	209	3723

Note: CL: Calpain sequences. SB: Curated Calpain Subtrate sequences. XSB: Computationally Expanded Substrate sequences. CT: Calpastatin sequences.

We strove to experimentally validate all SB cleavage sites: out of the collected substrates, we generated 94 sequences with a length of 20 centered around the cleavage site, then biochemically evaluated cleavage efficiency using purified calpains and iTRAQ^{TM} labeling. Each iTRAQ^{TM} signal value gives a coarse estimate of proteolysis efficiency of the corresponding peptides by calpains. The iTRAQ^{TM} signal is shown in the corresponding substrate entry for 56 cases in which cleavage sites were experimentally confirmed. Fig. 1 (d) shows a sample screenshot from the 'Substrates' category, including the iTRAQ^{TM} signal field.

In addition, we computationally expanded the set of SB sequences to 1,914 putative substrates (XSB) using BLAST to find similar sequences in Entrez: More specifically, for each (query) sequence (QS) in SB, we cut out a peptide (QP) of 60 amino acids centered around the cleavage site to cast a query. We then selected the returned sequences as XSB if they satisfied all three following conditions:

(1) The returned sequence could be aligned with QS with matching rate of over 90%.
(2) Pairwise alignment between the returned sequence and QS yielded a local alignment score of over 1,000.
(3) The corresponding cleavage site of the returned sequence could be aligned with QP with a matching rate of over 90%.

For 'Calpastatin', we collected 209 sequences by searching for entries in Entrez which include *Calpain_inhib* anywhere in the entry and *calpastatin* in the definition line.

3. Cleavage Site Prediction

A notable consequence of the efforts that went into building the 'Substrates' section of CaMPDB, is the resulting availability of a large dataset of experimentally verified calpain substrate sequences, fully labeled with cleavage information. Allowing us for the first time to train efficient cleavage prediction tools with supervised learning methods.

The 'Predict' tab of the 'Substrates' section provides a prediction tool that will output position-based scores reflecting the probability of cleavage by calpain for any given sequence. This tool offers three separate prediction model options based on PSSM and Support Vector Machines (SVM) algorithms with either linear or Radial Basis Function (RBF) kernels.

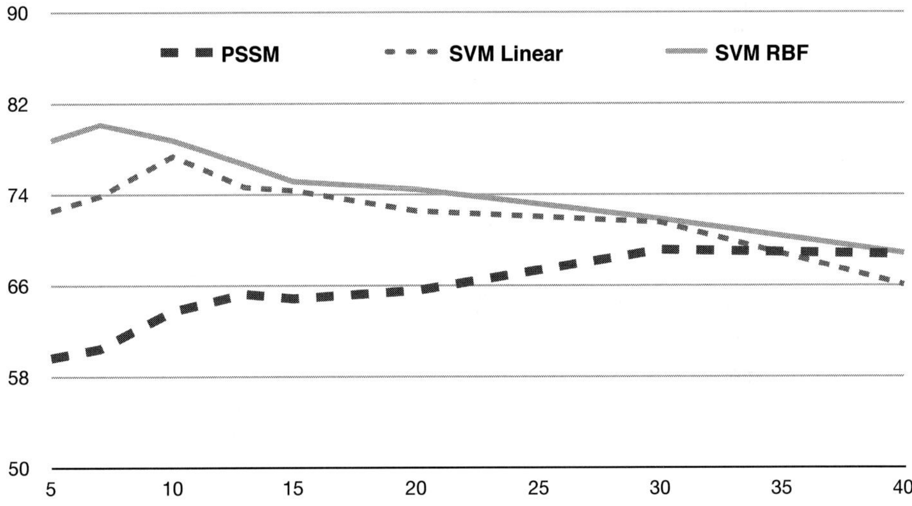

Fig. 3: AUC scores (in %) for each model (PSSM, SVM with linear kernel and SVM with RBF kernel), function of the number of amino acids used on each side of the putative cleavage site.

The PSSM model examines subsequences of length L centered around putative cleavage sites to compute the log-likelihood of cleavage, given the set of 267 known sites (extracted from the 104 SB sequences). Summing PSSM log-values, such as those under the 'Statistics' section, for each position in a window of length L yields a single score for this position. By sliding this window of length L over the query sequence it is possible to provide scores for the entire sequence.

The SVM model uses a similar sliding-window approach: a position-based binary vector of features is extracted for each sub-sequence of length L and scored using an SVM classifier with linear or RBF kernel. In addition to using the same set of

267 positive instances as the PSSM model, a random sample of negative instances is also selected out of all non-cleaved position in SB sequences.

Top performances for each method were measured using Area Under the ROC Curve (AUC) with 10x10-fold cross-validation and reached maximal values of 69.1% (SEM: 0.75%) for PSSM ($L = 2x30$), 77.3% (SEM: 0.04%) for SVM linear and 80.1% (SEM: 0.04%) for SVM RBF (using respectively $L = 2x7$ and $L = 2x10$ amino acids on each side).

Performances for a wider range of window lengths have been reported in Fig. 3. While it should be noted that PSSM performances for higher window lengths are likely due to some amount of overfitting (unavoidable given the size of the training set, despite the use of stringent cross-validation standards), the overall results provide a good baseline to which other, more sophisticated, supervised learning methods can be compared. With best prediction scores obtained within 10 amino acids of the cleavage site for SVM-based predictors, it seems possible to infer some early hypotheses regarding amino acid position specificity. Furthermore, significantly better results obtained with RBF over linear kernel, for smaller window sizes (between 5 and 10 amino acids on each side), would suggest strong non-linear correlations between some amino acid positions and cleavage selection. Finally, the study of results for varying lengths of extension around cleavage sites reveals some interesting asymmetry between left and right side (see Fig. 4): for both RBF and linear kernel SVM predictors, statistically better performances are achieved when focusing on the right (primed) side of the cleavage site.

Based on empirical evidence and performances during cross-validation testing, the value of L for the online tool was set to 40 amino acids (20 on each side of the cleavage site) for PSSM and 20 (2x10) for SVM. These values are justified by

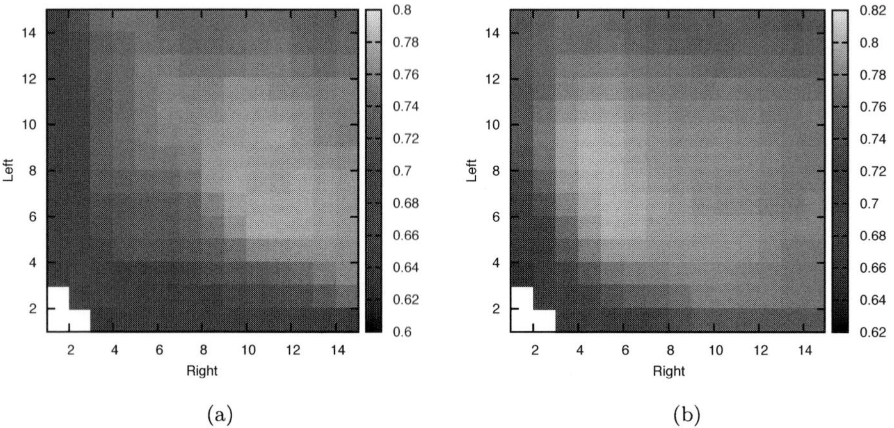

(a)   (b)

Fig. 4: AUC results (with 10x10-fold cross-validation) for SVM with Linear (a) and RBF (b) kernel, as a function of length of amino acids used on left and right side of cleavage site ($SEM < 0.0006$ for all values).

Cleavage Site Predictions

Sequence: *H2B histone family, member L [Bos taurus]*

10 Best scores:

1. Pos. **116** - Score: 0.20
2. Pos. **53** - Score: 0.19
3. Pos. **20** - Score: 0.15
4. Pos. **46** - Score: 0.14
5. Pos. **64** - Score: 0.13
6. Pos. **101** - Score: 0.12
7. Pos. **81** - Score: 0.12
8. Pos. **16** - Score: 0.11
9. Pos. **106** - Score: 0.10
10. Pos. **38** - Score: 0.10

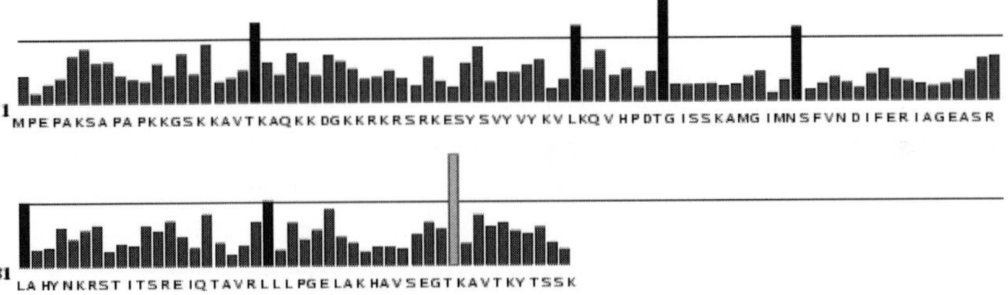

Fig. 5: Sample screen-capture of the cleavage prediction tool.

the 3D structure of a calpain/calpastatin complex [4, 8]. Calpastatin firmly binds to the calpain protease domain by approximately 20 residues as shown in Fig. 6. In other words, 2x10 residues are in close contact with calpain, and, thus, involved in determination of specificity. These values are also consistent with the approach taken by Ref. [17] to predict cleavage sites of a calpain substrate.

The online tool returns its results as a list of top cleavage site candidates and a bar chart presentation of scores along the input sequence (see Fig. 5).

4. Conclusion and Outlook

With CaMPDB, we provided a one-stop repository for all calpain and calpain-related information. In addition to richly annotated sequence entries and database-wide statistics, we also made available a useful first-approach tool to identify potential calpain substrates and cleavage sites.

Future directions for this project include the use of other machine-learning techniques (Markovian models and Bayesian networks) in order to improve the interpretability of results and bring a better understanding of the underlying biological model.

Acknowledgements

The authors would like to thank Suguru Koyama, Yumiko Kadoyama, Chikako Hayashi, Shoji Hata, Koichi Ojima and Yoshifumi Matsushima for developing

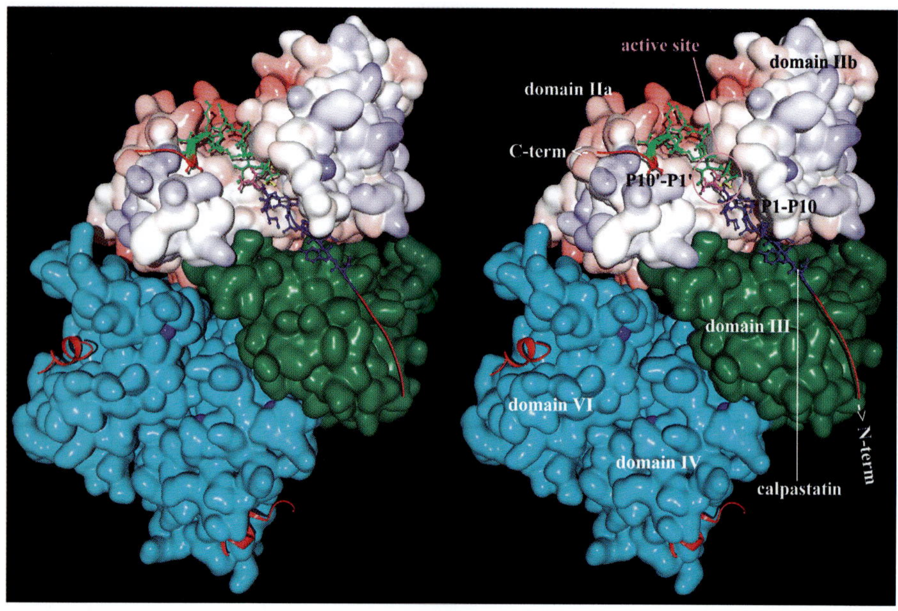

Fig. 6: Cross-eye stereo 3D structure of a rat m-calpain/calpastatin complex. Figures are drawn with FiatLux MolFeat Ver.4.0 using PDB data, 3DF0. The structure of m-calpain is shown by surface representation; domain III and two penta-EF-hand (PEF) domains (domains IV and VI) are indicated in green and blue, respectively. The protease domain (sub-domains IIa+IIb) is colored according to the electrostatic potential (red, - (acidic); blue, + (basic)). Active site residues, Cys105, His262, and Asn286 (NP_058812) in the protease domain are indicated by ball-stick drawing in yellow and a pink circle. Blue balls (visible in domains IV and VI) represent Ca2+. Calpastatin reactive site residues bound to domains IIa, IIb, and III of m-calpain are shown by ball-stick drawing (green, pink, and blue denote P10-P1, Glu236 (P27321) at the center, P1-P10 residues, respectively). Note that calpastatin, unlike calpain substrates, loops out the Glu236 residue at the center of the peptide. This is caused by the adjacent Gly237 residue, so that calpastatin is not immediately proteolyzed by calpains even though it firmly binds to the calpain protease domain, like regular substrates. Calpastatin also binds to two PEF domains on both sides of the reactive site, and, as a whole, it strongly inhibits calpain activity in a very specific manner.

the database. This work has been supported in part by BIRD and CREST of Japan Science and Technology Agency (JST), by MEXT.KAKENHI 18076007, by JSPS.KAKENHI 19658057, 20370055, and by a Takeda Science Foundation research grant.

Appendix A. Methods for iTRAQ quantification

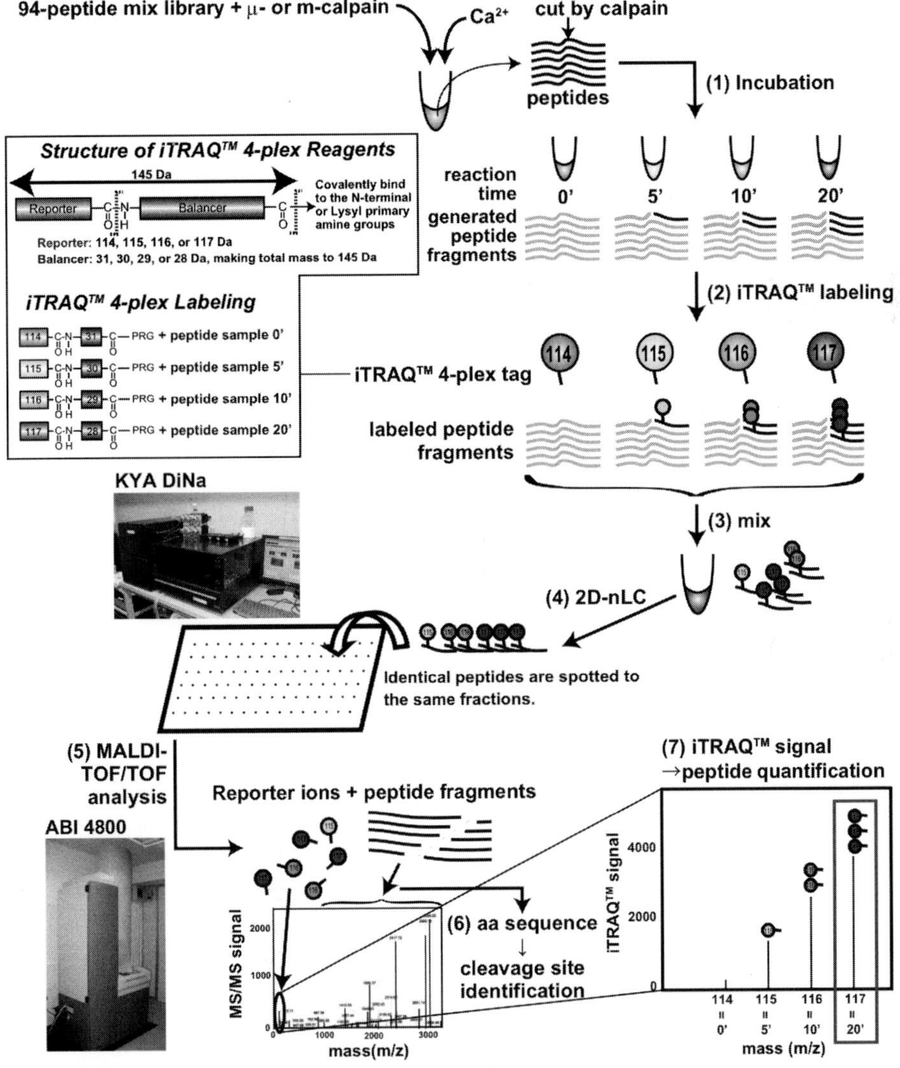

Fig. A1: A schematic figure of the iTRAQ quantification procedure.

Qualitative evaluation of the proper proteolytic cleavage of each peptide was performed using mass-spectrometry and iTRAQ^{TM} (isobaric **T**ag for **R**elative and **A**bsolute **Q**uantitation) methods [18], which will be described in the following (More detailed procedures and results of which will be published elsewhere.):

Briefly, peptides corresponding to calpain substrates (20mer, cleavage sites in the middle) were synthesized, mixed equally in molar quantity, and were digested with recombinant calpains prepared previously [11] (Fig. A1 (1)). At various time points, the reaction mixture was labeled with iTRAQ^{TM} 4-plex reagents, which are composed of "reporter" (Mr: 114 117 Da) and "balancer" (Mr: 31 28 Da, respectively) to make their total Mr isobaric (Fig. A1 (2)). All samples from various time points were, then mixed (Fig. A1 (3)), separated by 2D-nLC (**2-D**imensional **n**ano-scale **L**iquid-**C**hromatography; 1st step: strong cation ion-exchange chromatography; 2nd step: reverse-phase chromatography) into 684 fractions [10] (Fig. A1 (4)). Each fraction was automatically mixed with -cyano-4-hydroxycinnamic acid containing matrix solution, and spotted onto ABI 4800 MALDI target plates. Fractionated peptides were detected by MALDI-TOF/TOF analysis (Fig. A1 (5)), which revealed the amino acid sequence of the peptide (Fig. A1 (6)) as well as relative amounts of the peptide at various time points (Fig. A1 (7)). For example, FKAKKAAMMT, which corresponds to Ryanodine receptor (SB0015) proteolyzed by calpain at the center [19], was increasingly detected incubation time-dependently, showing an iTRAQ^{TM}-117 signal of 4310.9. Although iTRAQ^{TM} signals detected by MALDI-TOF/TOF system are not quantitatively precise and rather qualitative with regard to amount comparison between different peptides: strong signals tend to indicate abundant existence of the peptides. Thus, iTRAQ^{TM} signal at the 20 min time point, if detected, are shown in the database with their evaluation compared to other detected signals (1 to 5 stars) The number of stars indicates qualitatively how effectively the peptide is proteolyzed at the indicated site.

References

[1] Bertipaglia, L., Carafoli, E., Calpains and human disease. *Subcell Biochem.*, 45:29–53, 2007.

[2] Friedrich, P., Bozoky, Z., Digestive versus regulatory proteases: on calpain action *in vivo*. *Biol. Chem.*, 386:609–612, 2005.

[3] Goll, D.E., et al., The calpain system. *Physiol Rev.*, 83:731–801, 2003.

[4] Hanna, R.A., Campbell, R.L., Davies, P.L., Calcium-bound structure of calpain and its mechanism of inhibition by calpastatin. *Nature*, 456(7220):409–412, 2008.

[5] Henikoff, S., Henikoff, J.G., Position-based sequence weights. *JMB*, 243:574–578, 1994.

[6] Igarashi, Y., et al., CutDB: a proteolytic event database. *NAR*, 35:D546–D549, 2007.

[7] Mizushima, N., Autophagy: process and function. *Genes Dev.*, 21:2861–2873, 2007.

[8] Moldoveanu, T., Gehring, K., Green, D.R., Concerted multi-pronged attack by calpastatin to occlude the catalytic cleft of heterodimeric calpains. *Nature*, 456(7220):404–408, 2008.

[9] Molinari, M., Anagli, J., Carafoli, E., PEST sequences do not influence substrate susceptibility to calpain proteolysis. *J. Biol. Chem.*, 270:2032–2035, 1995.

[10] Ono, Y., Hayashi, C., Doi, N., Kitamura, F., Shindo, M., Kudo, K., Tsubata, T., Yanagida, M., Sorimachi, H., Comprehensive survey of p94/calpain 3 substrates by comparative proteomics-Possible regulation of protein synthesis by p94. *Biotechnology Journal*, 2(5):565-576, 2007.

[11] Ono,Y., Kakinuma, K., Torii, F., Irie, A., Nakagawa, K., Labeit, S., Abe, K., Suzuki, K., Sorimachi, H., Possible regulation of the conventional calpain system by skeletal muscle-specific calpain, p94/calpain 3. *Journal of Biological Chemistry*, 279(4):2761-2771, 2004.

[12] Rawlings, N.D. et al. MEROPS: the peptidase database. *NAR*, 36:D320–D325, 2008.

[13] Rechsteiner, M., PEST sequences are signals for rapid intracellular proteolysis. *Semin. Cell Biol.*, 1:433–440, 1990.

[14] Schneider, T.D., Stephens, R.M., Sequence logos: A new way to display consensus sequences. *NAR*, 18:6097–6100, 1990.

[15] Suzuki, K., et al. Structure, activation, and biology of calpain.digestive versus regulatory proteases: on calpain action *in vivo*. *Diabetes*, 386:S12–S18, 2005.

[16] Tanaka, K., The proteasome: overview of structure and functions. *Proc. Jpn. Acad. Ser. B Phys. Biol. Sci.*, 85:12–36, 2009.

[17] Tompa, P., et al. On the sequential determinants of calpain cleavage. *J. Biol. Chem.*, 279(20):20775–20785, 2004.

[18] Zieske, L.R., A perspective on the use of iTRAQ^{TM} reagent technology for protein complex and profiling studies. *Journal of Experimental Botany*, 57(7):1501–1508, 2006.

[19] Zimmerman, U.J., Schlaepfer, W.W., Two-stage autolysis of the catalytic subunit initiates activation of calpain I. *Biochimica et Biophysica Acta*, 1078(2):192–198, 1991.

AUTHOR INDEX